Keyword Search in Databases

Synthesis Lectures on Data Management

Editor
M. Tamer Özsu

Keyword Search in Databases
Jeffrey Xu Yu, Lu Qin, and Lijun Chang
2010

Keyword Search in Databases

Jeffrey Xu Yu, Lu Qin, and Lijun Chang

ISBN: 978-3-031-79425-4   paperback
ISBN: 978-3-031-79426-1   ebook

DOI 10.1007/978-3-031-79426-1

A Publication in the Springer series
SYNTHESIS LECTURES ON DATA MANAGEMENT

Lecture #1
Series Editor: M. Tamer Özsu
Series ISSN
Synthesis Lectures on Data Management
ISSN pending.

Keyword Search in Databases

Jeffrey Xu Yu, Lu Qin, and Lijun Chang
Chinese University of Hong Kong

SYNTHESIS LECTURES ON DATA MANAGEMENT #1

ABSTRACT

It has become highly desirable to provide users with flexible ways to query/search information over databases as simple as keyword search like Google search.

This book surveys the recent developments on keyword search over databases, and focuses on finding structural information among objects in a database using a set of keywords. Such structural information to be returned can be either trees or subgraphs representing how the objects, that contain the required keywords, are interconnected in a relational database or in an XML database. The structural keyword search is completely different from finding documents that contain all the user-given keywords. The former focuses on the interconnected object structures, whereas the latter focuses on the object content.

The book is organized as follows. In Chapter 1, we highlight the main research issues on the structural keyword search in different contexts. In Chapter 2, we focus on supporting structural keyword search in a relational database management system using the SQL query language. We concentrate on how to generate a set of SQL queries that can find all the structural information among records in a relational database completely, and how to evaluate the generated set of SQL queries efficiently. In Chapter 3, we discuss graph algorithms for structural keyword search by treating an entire relational database as a large data graph. In Chapter 4, we discuss structural keyword search in a large tree-structured XML database. In Chapter 5, we highlight several interesting research issues regarding keyword search on databases.

The book can be used as either an extended survey for people who are interested in the structural keyword search or a reference book for a postgraduate course on the related topics.

KEYWORDS

keyword search, interconnected object structures, relational databases, XML databases, data stream, rank

Contents

Preface

It has become highly desirable to provide flexible ways for users to query/search information by integrating database (DB) and information retrieval (IR) techniques in the same platform. On one hand, the sophisticated DB facilities provided by a database management system assist users to query well-structured information using a query language based on database schemas. Such systems include conventional RDBMSs (such as *DB2*, *ORACLE*, SQL-Server), which use SQL to query relational databases (*RDBs*) and *XML* data management systems, which use XQuery to query *XML* databases. On the other hand, IR techniques allow users to search unstructured information using keywords based on scoring and ranking, and they do not need users to understand any database schemas. The main research issues on DB/IR integration are discussed by Chaudhuri et al. [2005] and debated in a SIGMOD panel discussion [Amer-Yahia et al., 2005]. Several tutorials are also given on keyword search over *RDBs* and *XML* databases, including those by Amer-Yahia and Shanmugasundaram [2005]; Chaudhuri and Das [2009]; Chen et al. [2009].

The main purpose of this book is to survey the recent developments on keyword search over databases that focuses on finding *structural* information among objects in a database using a keyword query that is a set of keywords. Such structural information to be returned can be either trees or sub-graphs representing how the objects, which contain the required keywords, are interconnected in an *RDB* or in an *XML* database. In this book, we call this *structural keyword search* or, simply, *keyword search*. The structural keyword search is completely different from finding documents that contain all the user-given keywords. The former focuses on the interconnected object structures, whereas the latter focuses on the object content. In a DB/IR context, for this book, we use keyword search and keyword query interchangeably. We introduce forms of answers, scoring/ranking functions, and approaches to process keyword queries.

The book is organized as follows.

In Chapter 1, we highlight the main research issues on the structural keyword search in different contexts.

In Chapter 2, we focus on supporting keyword search in an RDMS using SQL. Since this implies making use of the database schema information to issue SQL queries in order to find structural information for a keyword query, it is generally called a schema-based approach. We concentrate on the two main steps in the schema-based approach, namely, how to generate a set of SQL queries that can find all the structural information among tuples in an *RDB* completely and how to evaluate the generated set of SQL queries efficiently. We will address how to find all or top-k answers in a static *RDB* or a dynamic data stream environment.

In Chapter 3, we also focus on supporting keyword search in an RDBMS. Unlike the approaches discussed in Chapter 2 using SQL, we discuss the approaches that are based on graph algorithms by materializing an entire database as a large data graph. This type of approach is called schema-free, in the sense that it does not request any database schema assistance. We introduce several algorithms, namely polynomial delay based algorithms, dynamic programming based algorithms, and Dijkstra shortest path based algorithms. We discuss how to find exact top-k and approximate top-k answers in a large data graph for a keyword query. We will discuss the indexing mechanisms and the ways to handle a large graph on disk.

In Chapter 4, we discuss keyword search in an *XML* database where an *XML* database is a large data tree. The two main issues are how to find all subtrees that contain all the user-given keywords and how to identify the meaning of such returned subtrees. We will discuss several algorithms to find subtrees based on lowest common ancestor (LCA) semantics, smallest LCA semantics, exclusive LCA semantics, etc.

In Chapter 5, we highlight several interesting research issues regarding keyword search on databases. The topics include how to select a database among many possible databases to answer a keyword query, how to support keyword query in a spatial database, how to rank objects according to their relevance to a keyword query using PageRank-like approaches, how to process keyword queries in an OLAP (On-Line Analytical Processing) context, how to find frequent additional keywords that are most related to a keyword query, how to interpret a keyword query by showing top-k SQL queries, and how to project a small database that only contains objects related to a keyword query.

The book surveys the recent developments on the structural keyword search. The book can be used as either an extended survey for people who are interested in the structural keyword search or a reference book for a postgraduate course on the related topics.

We acknowledge the support of our research on keyword search by the grant of the Research Grants Council of the Hong Kong SAR, China, No. 419109.

We are greatly indebted to M. Tamer Özsu who encouraged us to write this book and provided many valuable comments to improve the quality of the book.

Jeffrey Xu Yu, Lu Qin, and Lijun Chang
The Department of Systems Engineering and Engineering Management
The Faculty of Engineering
The Chinese University of Hong Kong
December, 2009

CHAPTER 1

Introduction

Conceptually, a database can be viewed as a data graph $G_D(V, E)$, where V represents a set of objects, and E represents a set of connections between objects. In this book, we concentrate on two kinds of databases, a relational database (RDB) and an XML database. In an RDB, an object is a tuple that consists of many attribute values where some attribute values are strings or full-text; there is a connection between two objects if there exists at least one reference from one to the other. In an XML database, an object is an element that may have attributes/values. Like RDBs, some values are strings. There is a connection (parent/child relationship) between two objects if one links to the other. An RDB is viewed as a large graph, whereas an XML database is viewed as a large tree.

The main purpose of this book is to survey the recent developments on finding *structural information* among objects in a database using a keyword query, Q, which is a set of keywords of size l, denoted as $Q = \{k_1, k_2, \cdots, k_l\}$. We call it an l-keyword query. The structural information to be returned for an l-keyword query can be a set of connected structures, $\mathcal{R} = \{R_1(V, E), R_2(V, E), \cdots\}$ where $R_i(V, E)$ is a connected structure that represents how the objects that contain the required keywords, are interconnected in a database G_D. S can be either all trees or all subgraphs. When a function $score(\cdot)$ is given to score a structure, we can find the top-k structures instead of all structures in the database G_D. Such a $score(\cdot)$ function can be based on either the text information maintained in objects (node weights) or the connections among objects (edge weights), or both.

In Chapter 2, we focus on supporting keyword search in an RDBMS using SQL. Since this implies making use of the database schema information to issue SQL queries in order to find structures for an l-keyword query, it is called the schema-based approach. The two main steps in the schema-based approach are how to generate a set of SQL queries that can find all the structures among tuples in an RDB completely and how to evaluate the generated set of SQL queries efficiently. Due to the nature of set operations used in SQL and the underneath relational algebra, a data graph G_D is considered as an undirected graph by ignoring the direction of references between tuples, and, therefore, a returned structure is of undirected structure (either tree or subgraph). The existing algorithms use a parameter to control the maximum size of a structure allowed. Such a size control parameter limits the number of SQL queries to be executed. Otherwise, the number of SQL queries to be executed for finding all or even top-k structures is too large. The $score(\cdot)$ functions used to rank the structures are all based on the text information on objects. We will address how to find all or top-k structures in a static RDB or a dynamic data stream environment.

In Chapter 3, we focus on supporting keyword search in an RDBMS from a different viewpoint, by treating an RDB as a directed graph G_D. Unlike an undirected graph, the fact that an object v can reach to another object u in a directed graph does not necessarily mean that the object v is reachable from u. In this context, a returned structure (either steiner tree, distinct rooted tree, r-radius steiner graph, or multi-center subgraph) is directed. Such direction handling provides users with more information on how the objects are interconnected. On the other hand, it requests higher computational cost to find such structures. Many graph-based algorithms are designed to find top-k structures, where the $score(\cdot)$ functions used to rank the structures are mainly based on the connections among objects. This type of approach is called schema-free in the sense that it does not request any database schema assistance. In this chapter, we introduce several algorithms, namely polynomial delay based algorithms, dynamic programming based algorithms, and Dijkstra shortest path based algorithms. We discuss how to find exact top-k and approximate top-k structures in G_D for an l-keyword query. The size control parameter is not always needed in this type of approach. For example, the algorithms that find the optimal top-k steiner trees attempt to find the optimal top-k steiner trees among all possible combinations in G_D without a size control parameter. We also discuss the indexing mechanisms and the ways to handle a large graph on disk.

In Chapter 4, we discuss keyword search in an *XML* database where an *XML* database is considered as a large directed tree. Therefore, in this context, the data graph G_D is a directed tree. Such a directed tree may be seen as a special case of the directed graph, so that the algorithms discussed in Chapter 3 can be used to support l-keyword queries in an *XML* database. However, the main research issue is different. The existing approaches process l-keyword queries in the context of *XML* databases by finding structures that are based on the lowest common ancestor (LCA) of the objects that contain the required keywords. In other words, a returned structure is a subtree rooted at the LCA in G_D that contains the required keywords in the subtree, but it is not any subtree in G_D that contains the required keywords in the subtree. The main research issue is to efficiently find meaningful structures to be returned. The meaningfulness are not defined based on $score(\cdot)$ functions. Algorithms are proposed to find smallest LCA, exclusive LCA, and compact LCA, which we will discuss in Chapter 4.

In Chapter 5, we highlight several interesting research issues regarding keyword search on databases. The topics include how to select a database among many possible databases to answer an l-keyword query, how to support l-keyword queries in a spatial database, how to rank objects according to their relevance to an l-keyword query using PageRank-like approaches, how to process l-keyword queries in an OLAP (On-Line Analytical Processing) context, how to find frequent additional keywords that are most related to an l-keyword query, how to interpret an l-keyword query by showing top-k SQL queries, and how to project a small database that only contains objects related to an l-keyword query.

CHAPTER 2

Schema-Based Keyword Search on Relational Databases

In this chapter, we discuss how to support keyword queries in a middleware on top of a RDBMS or on a RDBMS directly using SQL. In Section 2.1, we start with fundamental definitions such as, a schema graph, an l-keyword query, a tree-structured answer that is called a minimal total joining network of tuples and is denoted as *MTJNT*, and ranking functions. In Section 2.2, for evaluating an l-keyword query over an *RDB*, we discuss how to generate query plans (called candidate network generation), and in Section 2.3, we discuss how to evaluate query plans (called candidate evaluation). In particular, we discuss how to find all *MTJNT*s in a static *RDB* and a dynamic *RDB* in a data stream context, and we discuss how to find top-k *MTJNT*s. In Section 2.4, in addition to the tree-structured answers (*MTJNT*s) to be found, we discuss how to find graph structured answers using SQL on RDBMS directly.

2.1 INTRODUCTION

We consider a relational database schema as a directed graph $G_S(V, E)$, called a *schema graph*, where V represents the set of relation schemas $\{R_1, R_2, \cdots, R_n\}$ and E represents the set of edges between two relation schemas. Given two relation schemas, R_i and R_j, there exists an edge in the schema graph, from R_i to R_j, denoted $R_i \rightarrow R_j$, if the primary key defined on R_i is referenced by the foreign key defined on R_j. There may exist multiple edges from R_i to R_j in G_S if there are different foreign keys defined on R_j referencing the primary key defined on R_i. In such a case, we use $R_i \xrightarrow{X} R_j$, where X is the foreign key attribute names. We use $V(G_S)$ and $E(G_S)$ to denote the set of nodes and the set of edges of G_S, respectively. In a relation schema R_i, we call an attribute, defined on strings or full-text, a text attribute, to which keyword search is allowed.

A relation on relation schema R_i is an instance of the relation schema (a set of tuples) conforming to the relation schema, denoted $r(R_i)$. We use R_i to denote $r(R_i)$ if the context is obvious. A relational database (*RDB*) is a collection of relations. We assume, for a relation schema, R_i, there is an attribute called TID (Tuple ID), a tuple in $r(R_i)$ is uniquely identified by a TID value in the entire *RDB*. In *ORACLE*, a hidden attribute called rowid in a relation can be used to identify a tuple in an *RDB*, uniquely. In addition, such a TID attribute can be easily supported as a composite attribute in a relation, R_i, using two attributes, namely, relation-identifier and tuple-identifier. The former keeps the unique relation schema identifier for R_i, and the latter keeps a unique tuple identifier in

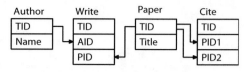

Figure 2.1: *DBLP* Database Schema [Qin et al., 2009a]

relation $r(R_i)$. Together with the two values, a tuple is uniquely identified in the entire *RDB*. For simplicity and without loss of generality, in the following discussions, we assume primary keys are TID, and we use primary key and TID interchangeably.

Given an *RDB* on the schema graph, G_S, we say two tuples t_i and t_j in an *RDB* are connected if there exists at least one foreign key reference from t_i to t_j or vice versa, and we say two tuples t_i and t_j in an *RDB* are reachable if there exists at least a sequence of connections between t_i and t_j. The distance between two tuples, t_i and t_j, denoted as $dis(t_i, t_j)$, is defined as the minimum number of connections between t_i and t_j.

An *RDB* can be viewed as a database graph $G_D(V, E)$ on the schema graph G_S. Here, V represents a set of tuples, and E represents a set of connections between tuples. There is a connection between two tuples, t_i and t_j in G_D, if there exists at least one foreign key reference from t_i to t_j or vice versa (undirected) in the *RDB*. In general, two tuples, t_i and t_j are reachable if there exists a sequence of connections between t_i and t_j in G_D. The distance $dis(t_i, t_j)$ between two tuples t_i and t_j is defined the same as over an *RDB*. It is worth noting that we use G_D to explain the semantics of keyword search but do not materialize G_D over *RDB*.

Example 2.1   A simple *DBLP* database schema, G_S, is shown in Figure 2.1. It consists of four relation schemas: `Author`, `Write`, `Paper`, and `Cite`. Each relation has a primary key TID. `Author` has a text attribute `Name`. `Paper` has a text attribute `Title`. `Write` has two foreign key references: `AID` (refers to the primary key defined on `Author`) and `PID` (refers to the primary key defined on `Paper`). `Cite` specifies a citation relationship between two papers using two foreign key references, namely, `PID1` and `PID2` (paper `PID2` is cited by paper `PID1`), and both refer to the primary key defined on `Paper`. A simple *DBLP* database is shown in Figure 2.2. Figure 2.2(a)-(d) show the four relations, where x_i means a primary key (or TID) value for the tuple identified with number i in relation x (a, p, c, and w refer to `Author`, `Paper`, `Cite`, and `Write`, respectively). Figure 2.2(e) illustrates the database graph G_D for the simple *DBLP* database. The distance between a_1 and p_1, $dis(a_1, p_1)$, is 2.

An l-keyword query is given as a set of keywords of size l, $Q = \{k_1, k_2, \cdots, k_l\}$, and searches interconnected tuples that contain the given keywords, where a tuple contains a keyword if a text attribute of the tuple contains the keyword. To select all tuples from a relation R that contain a keyword k_1, a predicate $contain(A, k_1)$ is supported in sqL in IBM *DB2*, *ORACLE*, and Microsoft SQL-Server, where A is a text attribute in R. The following sqL query, finds all tuples in R containing

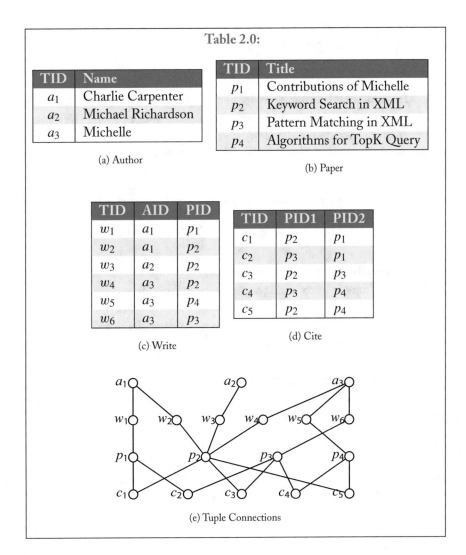

Figure 2.2: *DBLP* Database [Qin et al., 2009a]

k_1 provided that the attributes A_1 and A_2 are all and the only text attributes in relation R. We say a tuple contains a keyword, for example k_1, if the tuple is included in the result of such a selection.

select * **from** R **where** $contain(A_1, k_1)$ **or** $contain(A_2, k_1)$

An l-keyword query returns a set of answers, where an answer is a minimal total joining network of tuples (*MTJNT*) [Agrawal et al., 2002; Hristidis and Papakonstantinou, 2002] that is defined as follows.

Definition 2.2   Minimal Total Joining Network of Tuples (*MTJNT*).   Given an l-keyword query and a relational database with schema graph G_S, a joining network of tuples (*JNT*) is a connected tree of tuples where every two adjacent tuples, $t_i \in r(R_i)$ and $t_j \in r(R_j)$ can be joined based on the foreign key reference defined on relational schema R_i and R_j in G_S (either $R_i \to R_j$ or $R_j \to R_i$). An *MTJNT* is a joining network of tuples that satisfy the following two conditions:

- Total: each keyword in the query must be contained in at least one tuple of the joining network.

- Minimal: a joining network of tuples is not total if any tuple is removed.

Because it is meaningless if two tuples in an *MTJNT* are too far away from each other, a size control parameter, Tmax, is introduced to specify the maximum number of tuples allowed in an *MTJNT*.

Given an *RDB* on the schema graph G_S, in order to generate all the *MTJNT*s for an l-keyword query, the keyword relation and Candidate Network (*CN*) are defined as follows.

Definition 2.3   Keyword Relation.   Given an l-keyword query Q and a relational database with schema graph G_S, a keyword relation $R_i\{K'\}$ is a subset of relation R_i containing tuples that only contain keywords $K'(\subseteq Q))$ and no other keywords, as defined below:

$$R_i\{K'\} = \{t | t \in r(R_i) \land \forall k \in K', t \text{ contains } k \land \forall k \in (K - K'), t \text{ does not contain } k\}$$

where K is the set of keywords in Q, i.e., $K = Q$. We also allow K' to be $\emptyset$. In such a situation, $R_i\{\}$ consists of tuples that do not contain any keywords in Q and is called an empty keyword relation.

Definition 2.4   Candidate Network.   Given an l-keyword query Q and a relational database with schema graph G_S, a candidate network (*CN*) is a connected tree of keyword relations where for every two adjacent keyword relations $R_i\{K_1\}$ and $R_j\{K_2\}$, we have $(R_i, R_j) \in E(G_S)$ or $(R_j, R_i) \in E(G_S)$. A candidate network must satisfy the following two conditions:

- Total: each keyword in the query must be contained in at least one keyword relation of the candidate network.

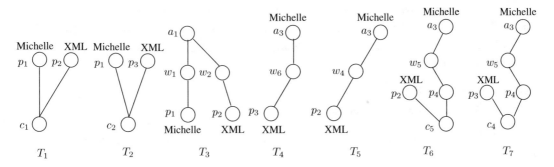

Figure 2.3: *MTJNT*s (Q = {Michelle, XML}, Tmax = 5)

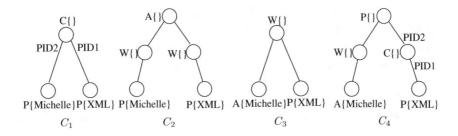

Figure 2.4: *CN*s (Q = {Michelle, XML}, Tmax = 5)

- Minimal: a candidate network is not total if any keyword relation is removed.

Generally speaking, a *CN* can produce a set of (possibly empty) *MTJNT*s, and it corresponds to a relational algebra that joins a sequence of relations to obtain *MTJNT*s over the relations involved. Given a keyword query Q and a relational database with schema graph G_S, let $\mathcal{C} = \{C_1, C_2, \cdots\}$ be the set of all candidate networks for Q over G_S, and let $\mathcal{T} = \{T_1, T_2, \cdots\}$ be the set of all *MTJNT*s for Q over the relational database. For every $T_i \in \mathcal{T}$, there is exactly one $C_j \in \mathcal{C}$ that produces T_i.

Example 2.5 For the *DBLP* database shown in Figure 2.2 and the schema graph shown in Figure 2.1. Suppose a 2-keyword query is Q = {Michelle, XML} and Tmax = 5. The seven *MTJNT*s are shown in Figure 2.3. The fourth one, $T_4 = a_3 \bowtie w_6 \bowtie p_3$, indicates that the author a_3 that contains the keyword "Michelle" writes a paper p_3 that contains the keyword "XML". The *JNT* $a_3 \bowtie w_5 \bowtie p_4$ is not an *MTJNT* because it does not contain the keyword "XML". The *JNT* $a_3 \bowtie w_6 \bowtie p_3 \bowtie c_2 \bowtie p_1$ is not an *MTJNT* because after removing tuples p_1 and c_2, it still contains all the keywords.

Four *CN*s are shown in Figure 2.4 for the keyword query $Q = \{$Michelle, XML$\}$ and Tmax $= 5$. P, C, W, and A represent the four relations, `Paper`, `Cite`, `Write`, and `Author`, in *DBLP* (Figure 2.1). The keyword relation $P\{$XML$\}$ means $\sigma_{contain(\text{XML})}(\sigma_{\neg contain(\text{Michelle})} P)$ or, equivalently, the following SQL query

select * **from** Paper **as** P
where contain(Title, XML) **and not** contain(Title, Michelle)

Note that there is only one text-attribute `Title` in the `Paper` relation. In a similar fashion, $P\{\}$ means

select * **from** Paper **as** P
where not contain(Title, XML) **and not** contain(Title, Michelle)

The first *CN* $C_1 = P\{$Michelle$\} \bowtie C\{\} \bowtie P\{XML\}$ can produce the two *MTJNT*s T_1 and T_2 as shown in Figure 2.3. The network $A\{$Michelle$\} \bowtie W\{\} \bowtie P\{$Michelle$\}$ is not a *CN* because it does not contain the keyword "XML". The network $P\{$Michelle, XML$\} \bowtie W\{\} \bowtie A\{$Michelle$\}$ is not a *CN* because after removing the keyword relations $W\{\}$ and $A\{$Michelle$\}$, it still contains all keywords.

For an *l*-keyword query over a relational database, the number of *MTJNT*s can be very large even if Tmax is small. It is ineffective to present users a huge number of results for a keyword query. In order to handle the effectiveness, for each *MTJNT*, T, for a keyword query Q, it also allows a score function $score(T, Q)$ defined on T in order to rank results. The top-k keyword query is defined as follows.

Definition 2.6   Top-k Keyword Query.   Given an *l*-keyword query Q, in a relational database, the top-k keyword query retrieves k *MTJNT*s $\mathcal{T} = \{T_1, T_2, ..., T_k\}$ such that for any two *MTJNT*s T and T' where $T \in \mathcal{T}$ and $T' \notin \mathcal{T}$, $score(T, Q) \leq score(T', Q)$.

Ranking issues for *MTJNT*s are discussed in many papers [Hristidis et al., 2003a; Liu et al., 2006; Luo et al., 2007]. They aim at designing effective ranking functions that capture both the textual information (e.g., IR-Styled ranking) and structural information (e.g., the size of the *MTJNT*) for an *MTJNT*. Generally speaking, there are two categories of ranking functions, namely, the attribute level ranking function and the tree level ranking function.

Attribute Level Ranking Function: Given an *MTJNT* T and a keyword query Q, the tuple level ranking function first assigns each text attribute for tuples in T an individual score and then combines them together to get the final score. *DISCOVER-II* [Hristidis et al., 2003a] proposed a score function as follows:

$$score(T, Q) = \frac{\sum_{a \in T} score(a, Q)}{size(T)} \tag{2.1}$$

Here $size(T)$ is the size of T, such as the number of tuples in T. Consider each text attribute for tuples in T as a virtual document, $score(a, Q)$ is the IR-style relevance score for the virtual document a in the $MTJNT$ T and Q that is defined as:

$$score(a, Q) = \sum_{k \in Q \cap a} \frac{1 + \ln(1 + \ln(tf(a,k)))}{(1-s) + s \cdot \frac{dl(a)}{avdl(Rel(a))}} \cdot idf(a,k) \qquad (2.2)$$

where

$$idf(a,k) = \ln(\frac{N(Rel(a))}{df(Rel(a),k) + 1}) \qquad (2.3)$$

$tf(a,k)$ is the number of appearances of keyword k in the text attribute a; $dl(a)$ is the length of the text attribute a; $Rel(a)$ is the relation the attribute a belongs to; $avdl(Rel(a))$ is the average length of text attributes in relation $Rel(a)$; s is a constant (usually 0.2), which controls the sensitivity of $dl(a)/avdl(Rel(a))$; $N(Rel(a))$ is the total number of tuples in the relation $Rel(a)$; and $df(Rel(a), k)$ is the number of tuples in $Rel(a)$ that contain keyword k.

The IR-style relevance score value for a single text attribute a and keyword query Q can be calculated using the full text search engine in a commercial RDBMS. For example, a query

select * **from** Paper **as** P
where contain(Title, XML) and contain(Title, Michelle)
order by score(1) **DESC**

returns the tuples of the `Paper` table that are relevant to the keyword query with "XML" and "Michelle" in the `Title` attribute, sorted by their IR-style relevance score values (Eq. 2.2).

Liu et al. [2006] propose a score function for an $MTJNT$ T and a keyword query Q that is defined as:

$$score(T, Q) = \sum_{k \in Q \cap T} score(Q, k) \cdot score(T, k) \qquad (2.4)$$

Here, $score(Q, k)$ is the importance of k in Q, and it can be simply assigned as the term frequency in the query. $score(T, k)$ can be calculated as follows.

$$score(T, k) = comb(score(a_1, k), score(a_2, k), \cdots) \qquad (2.5)$$

Here $a_1, a_2 \cdots$ are all text attributes of tuples in T, and

$$score(a, k) = \frac{ntf(a,k) \cdot idf(a,k)}{ndl(a) \cdot nsize(T)} \qquad (2.6)$$

where

$$ntf(a,k) = 1 + \ln(1 + \ln(tf(a,k))) \qquad (2.7)$$

and the following four normalizations are integrated:

- Tuple Tree Size Normalization:

$$nsize(T) = (1-s) + s \cdot \frac{size(T)}{avgsize} \qquad (2.8)$$

where $avgsize$ is the average size of all CNs for query Q.

- Document Length Normalization:

$$ndl(a) = ((1 - s) + s \cdot \frac{dl(a)}{avdl(Rel(a))}) \cdot (1 + ln(avdl(Rel(a)))) \qquad (2.9)$$

- Document Frequency Normalization: See Eq. 2.3.

- Inter-Document Weight Normalization:

$$comb(s_1, s_2, ...) = max(s_1, s_2, ...) \cdot (1 + \ln(1 + \ln(\frac{sum(s_1, s_2, ...)}{max(s_1, s_2, ...)}))) \qquad (2.10)$$

The aim of the normalization of any part is to prevent that part from being too large to become the dominant part of the score value.

Tree Level Ranking Function: In the attribute level ranking functions, each text attribute of an *MTJNT* is considered as a virtual document. Tree level ranking functions consider the whole *MTJNT* as a virtual document rather than each individual text attribute. Given an *MTJNT* T and a keyword query Q, the score function studied in *SPARK* [Luo et al., 2007] considers three scores: $score_a(T, Q)$, $score_b(T, Q)$ and $score_c(T, Q)$ where $score_a(T, Q)$ is called the TF-IDF score, $score_b(T, Q)$ is called the completeness score, and $score_c(T, Q)$ is called the size normalization score. The final score function is defined as:

$$score(T, Q) = score_a(T, Q) \cdot score_b(T, Q) \cdot score_c(T, Q) \qquad (2.11)$$

- TF-IDF score:

$$score_a(T, Q) = \sum_{k \in Q \cap T} \frac{1 + \ln(1 + \ln(tf(T, k)))}{(1 - s) + s \cdot \frac{dl(T)}{avdl(CN^*(T))}} \cdot \ln(idf(T, k)) \qquad (2.12)$$

where

$$tf(T, k) \quad = \quad \sum_{a \in T} tf(a, k) \qquad (2.13)$$

$$idf(T, k) \quad = \quad \frac{N(CN^*(T))}{df(CN^*(T), k) + 1} \qquad (2.14)$$

$tf(T, k)$ is the number of appearances of k in all text attributes for tuples in T; $dl(T)$ is the total length of all text attributes for tuples in T; $CN^*(T)$ is the joint relation of the *JNT* by removing the keyword selection prediction from each relation of the *CN* that the *MTJNT* T belongs to (for example, for the *MTJNT* $T = a_3 \bowtie w_6 \bowtie p_3$ from the *CN* $C = A\{Michelle\} \bowtie W\{\} \bowtie P\{XML\}$, we have $CN^*(T) = A\{\} \bowtie W\{\} \bowtie P\{\}$); $avdl(CN^*(T))$ is the average length of text attributes for all tuples in $CN^*(T)$; $N(CN^*(T))$ is the number of tuples in the relation $CN^*(T)$; and $df(CN^*(T), k)$ is the number of tuples in the relation $CN^*(T)$ that contain keyword k in the text attributes. Eq. 2.14 looks similar to Eq. 2.3, but it is used in the tree level ranking function instead of the attribute level ranking function.

- Completeness Score:

$$score_b(T, Q) = 1 - \left(\frac{\sum_{k \in Q}(1 - I(T, Q, k))^p}{l}\right)^{\frac{1}{p}} \qquad (2.15)$$

where $I(T, Q, k)$ is the normalized term frequency for keyword k in the *MTJNT* T, i.e.,

$$I(T, Q, k) = \frac{tf(T, k)}{max_{k_i \in Q} tf(T, k_i)} \cdot \frac{idf(T, k)}{max_{k_i \in Q} idf(T, k_i)} \qquad (2.16)$$

Consider the normalized term frequency for all keywords that form an l-dimensional vector space where values are in $[0, 1]$. When all keywords have the same non-zero frequency in the *MTJNT*, it will reach the ideal position $[1, 1, ...1]$. L_p distance between the current position and the ideal position is used to calculate the completeness score of the current *MTJNT*. $p \leq 1$ is a tuning factor; as p increases, a switch occurs from OR semantics to AND semantics[1]. For example, when $p \to +\infty$, the completeness score will become $min_{k \in Q} I(T, Q, k)$, which gives a 0 score to those *MTJNT*s that do not contain all the keywords. In the experiments performed on *SPARK* [Luo et al., 2007], $p = 2.0$ is shown to be good enough to get most of the answers under the AND semantics.

- Size Normalization Score:

$$score_c(T, Q) = (1 + s_1 - s_1 \cdot size(T)) \cdot (1 + s_2 - s_2 \cdot keysize(T, Q)) \qquad (2.17)$$

where $size(T)$ is the number of tuples in T; $keysize(T, Q)$ is the number of tuples in T that contain at least one keyword in Q; s_1 and s_2 are two parameters introduced to balance the sensitivities of $size(T)$ and $keysize(T, Q)$. In *SPARK* [Luo et al., 2007], s_1 is set to be 0.15 and s_2 is set to be $\frac{1}{|Q|+1}$. The size normalization score is similar as the tuple tree size normalization score in Eq. 2.8.

In the framework of RDBMS, the two main steps of processing an l-keyword query are candidate network generation and candidate network evaluation.

1. **Candidate Network Generation:** In the candidate network generation step, a set of candidate networks $\mathcal{C} = \{C_1, C_2, \cdots\}$ is generated over a graph schema G_S. The set of *CN*s shall be complete and duplication-free. The former ensures that all *MTJNT*s are found, and the latter is mainly for efficiency consideration.

2. **Candidate Network Evaluation:** In the candidate network evaluation step, all $C_i \in \mathcal{C}$ are evaluated according to different semantics/environments for the keyword query (e.g., the AND/OR semantics, the all/top-k semantics, the static/stream environments).

We will introduce the two steps one by one in the next two sections.

---

[1]In *SPARK* [Luo et al., 2007], each *MTJNT* is allowed to contain just a subset of keywords, i.e., the OR semantics.

2.2   CANDIDATE NETWORK GENERATION

In order to generate all candidate networks for an l-keyword query Q over a relational database with schema graph G_S, algorithms are designed to generate candidate networks $\mathcal{C} = \{C_1, C_2, ...\}$ that satisfy the following two conditions:

- **Complete:** For each solution T of the keyword query, there exists a candidate network $C_i \in \mathcal{C}$ that can produce T.

- **Duplication-Free:** For every two CNs $C_i \in \mathcal{C}$ and $C_j \in \mathcal{C}$, C_i and C_j are not isomorphic to each other.

The complete and duplication-free conditions ensure that (1) all results ($MTJNT$s) for a keyword query will be produced by the set of CNs generated (due to completeness), and (2) any result T for a keyword query will be produced only once, i.e., there does not exist two CNs $C_i \in \mathcal{C}$ and $C_j \in \mathcal{C}$ such that C_i and C_j both produce T (due to the duplication-free condition).

The first algorithm to generate all CNs was proposed in *DISCOVER* [Hristidis and Papakonstantinou, 2002]. It expands the partial CNs generated to larger partial CNs until all CNs are generated. As the number of partial CNs can be exponentially large, arbitrarily expanding will make the algorithm extremely inefficient. In *DISCOVER* [Hristidis and Papakonstantinou, 2002], there are three pruning rules for partial CNs.

- **Rule-1:** Duplicated CNs are pruned (based on tree isomorphism).

- **Rule-2:** A CN can be pruned if it contains all the keywords and there is a leaf node, $R_j\{K'\}$, where $K' = \emptyset$, because it will generate results that do not satisfy the condition of minimality.

- **Rule-3:** When there only exists a single foreign key reference between two relation schemas (for example, $R_i \rightarrow R_j$), CNs including $R_i\{K_1\} \rightarrow R_j\{K_2\} \leftarrow R_i\{K_3\}$ will be pruned, where K_1, K_2, and K_3 are three subsets of Q, and $R_i\{K_1\}, R_j\{K_2\}$, and $R_i\{K_3\}$ are keyword relations (refer to Definition 2.3).

The Rule-3 reflects the fact that the primary key defined on R_i and a tuple in the relation of $R_j\{K_2\}$ must refer to the same tuple appearing in both relations $R_i\{K_1\}$ and $R_i\{K_3\}$. As the same tuple cannot appear in two sub-relations in a CN (otherwise, it will not produce a valid $MTJNT$ because the minimal condition will not be satisfied), the join results for $R_i\{K_1\} \rightarrow R_j\{K_2\} \leftarrow R_i\{K_3\}$ will not contain any valid $MTJNT$.

The algorithm to generate all CNs in *DISCOVER* is shown in Algorithm 1. Lines 1-3 initialize a queue Q that maintains all partial CNs to be a list of trees with size 1 that contain any subset of keywords. From line 4, each partial tree T is dequeued from $\mathcal{Q}$, iteratively. Lines 6-7 check whether T is a valid CN that has not been generated before. If so, it is added to the result set $\mathcal{C}$ and there is no need to further expand T. Lines 8-13 expand T by adding an edge from any relation R_i in T to another new relation R_j, and form a new partial CN T'. T' is enqueued for further expansion if it does not satisfy the pruning rules 2 or 3.

Algorithm 1 Discover-CNGen (Q, Tmax, G_S)

Input: an l-keyword query $Q = \{k_1, k_2, \cdots, k_l\}$, the size control parameter Tmax,
      the schema graph G_S.
Output: the set of CNs $\mathcal{C} = \{C_1, C_2, \cdots\}$.

  1: $\mathcal{Q} \leftarrow \emptyset; \mathcal{C} \leftarrow \emptyset$
  2: **for all** $R_i \in V(G_S)$, $K' \subseteq Q$ **do**
  3:    $\mathcal{Q}.enqueue(R_i\{K'\})$
  4: **while** $\mathcal{Q} \neq \emptyset$ **do**
  5:    $T \leftarrow \mathcal{Q}.dequeue()$
  6:    **if** T is minimal and total **and** T does not satisfy Rule-1 **then**
  7:      $\mathcal{C} \leftarrow \mathcal{C} \bigcup \{T\}$; **continue**
  8:    **if** the size of $T < $ Tmax **then**
  9:      **for all** $R_i \in T$ **do**
 10:        **for all** $(R_i, R_j) \in E(G_S)$ **or** $(R_j, R_i) \in E(G_S)$ **do**
 11:          $T' \leftarrow T \bigcup (R_i, R_j)$
 12:          **if** T' does not satisfy Rule-2 or Rule-3 **then**
 13:            $\mathcal{Q}.enqueue(T')$
 14: **return** $\mathcal{C}$;

As an example, consider the 2-keyword query $Q = \{$Michelle, XML$\}$ and the database with schema graph shown in Figure 2.1. First, we enqueue the set of partial CNs with size 1. Then $T_1 = A\{$Michelle$\}$ will be enqueued. When expanding T_1, as there is an edge $A \rightarrow W$ in the schema graph, we can add the corresponding edge in T_1 and form another partial CN $T_2 = A\{$Michelle$\} \bowtie W\{\}$. When expanding T_2, we can add edges $W\{\} \leftarrow A\{$XML$\}$, $W\{\} \leftarrow P\{$Michelle$\}$ and $W\{\} \leftarrow P\{$XML$\}$, and obtain three partial trees $T_3 = A\{$Michelle$\} \bowtie W\{\} \bowtie A\{XML\}$, $T_4 = A\{$Michelle$\} \bowtie W\{\} \bowtie P\{$Michelle$\}$, and $T_5 = A\{$Michelle$\} \bowtie W\{\} \bowtie P\{XML\}$, respectively. T_3 satisfies Rule-3, and thus it will be pruned in line 12. T_4 is not a CN because it is not total. However T_4 does not satisfy any pruning condition, and thus will be enqueued in line 13 for further expansion. T_5 is already a valid $MTJNT$, and thus it will be added into the final result set $\mathcal{C}$ in line 7.

The above algorithm can generate a complete and duplication-free set of CNs, but the cost of generating the set of CNs is high. This is mainly due to the following three reasons:

- Given an l-keyword query and a large Tmax over a complex database schema G_S, the number of CNs to be generated can be very large. The main factors that make the problem challenging are the graph complexity (the number of nodes/edges in the database schema G_S, $|G_S|$), the number of keywords (l), and the maximum size of $MTJNT$ allowed (Tmax). The number of CNs increases exponentially while any $|G_S|$, l, or Tmax increases.

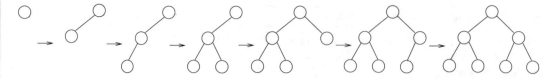

Figure 2.5: Rightmost path expansion

- The algorithm allows adding an arbitrary edge to an arbitrary position in a partial tree when expanding (line 9-13), which makes the number of temporal results extremely large, while only few of them will contribute to the final results. This is because most of the results will end up with a partial tree that is of size Tmax but does not contain all keywords (total). For example, for $\mathsf{Tmax} = 3$ and $Q = \{\text{Michelle, XML}\}$, over the database with schema graph shown in Figure 2.1, many will stop expansion in line 6 of Algorithm 1, such as $T = A\{\text{Michelle}\} \bowtie W\{\} \bowtie P\{\}$.

- The algorithm needs a large number of tree isomorphism tests, which is costly. This is because the isomorphism test will only be performed when a valid *MTJNT* is generated. As a result, all isomorphisms of an *MTJNT* will be generated and checked. For example, *MTJNT* $A\{\text{Michelle}\} \bowtie W\{\} \bowtie P\{\text{XML}\}$ can be generated through various ways such as $A\{\text{Michelle}\} \Rightarrow A\{\text{Michelle}\} \bowtie W\{\} \Rightarrow A\{\text{Michelle}\} \bowtie W\{\} \bowtie P\{\text{XML}\}$ and $P\{\text{XML}\} \Rightarrow W\{\} \bowtie P\{\text{XML}\} \Rightarrow A\{\text{Michelle}\} \bowtie W\{\} \bowtie P\{\text{XML}\}$.

In order to solve the above problems, *S-KWS* [Markowetz et al., 2007] proposes an algorithm (1) to reduce the number of partial results generated by expanding from part of the nodes in a partial tree and (2) to avoid isomorphism testing by assigning a proper expansion order. The solutions are based on the following properties:

- **Property-1:** For any partial tree, we can always find an expansion order, where every time, a new edge is added into the rightmost root-to-leaf path of the tree. An example for the rightmost expansion is shown in Figure 2.5, where a tree of size 7 is expanded by adding an edge to the rightmost path of the tree each time.

- **Property-2:** Every leaf node must contain a unique keyword if it is not on the rightmost root-to-leaf path of a partial tree. This is based on the rightmost path expansion discussed above. A leaf node which is not on the rightmost path of a partial tree will not be further expanded; in other words, it will be a leaf node of the final tree. If it does not contain a unique keyword, then we can simply remove it in order to satisfy the minimality of an *MTJNT*.

- **Property-3:** For any partial tree, we can always find a rightmost path expansion order, where the immediate subtrees of any node in the final expanded tree are lexicographically ordered. Actually, each subtree of a *CN* can be presented by an ordered string code. For example, for the CN $C_3 = A\{\text{Michelle}\} \bowtie W\{\} \bowtie P\{\text{XML}\}$ rooted at $W\{\}$ shown in Figure 2.4, it can be presented as either $W\{\}(A\{\text{Michelle}\})(P\{\text{XML}\})$ or $W\{\}(P\{\text{XML}\})(A\{\text{Michelle}\})$. The former is ordered while the latter is not ordered. We call the ordered string code the *canonical code* of the CN.

- **Property-4:** Even though the above order is fixed in expansion, the isomorphism cases may also happen because the CNs are un-rooted. The same CN may be generated multiple times by expansion from different roots that have different ordered string codes. To handle this problem, it needs to keep one which is lexicographically smallest among all ordered string codes (canonical codes) for the same CN. The smallest one can be used to uniquely identify the un-rooted CN.

- **Property-5:** Suppose the set of CNs is $\mathcal{C} = \{C_1, C_2, \cdots\}$. For any subset of keywords $K' \subseteq Q$ and any relation R, $\mathcal{C}$ can be divided into two parts $\mathcal{C}_1 = \{C_i | C_i \in \mathcal{C} \text{ and } C_i \text{ contain } R\{K'\}\}$ and $\mathcal{C}_2 = \{C_i | C_i \in \mathcal{C} \text{ and } C_i \text{ does not contain } R\{K'\}\}$. The two parts are disjoint and total. By disjoint, we mean that $\mathcal{C}_1 \bigcap \mathcal{C}_2 = \emptyset$ and by total, we mean that $\mathcal{C}_1 \bigcup \mathcal{C}_2 = \mathcal{C}$.

In order to make use of the above properties, the *expanded schema graph*, denoted G_X, is introduced. Given a relational database with schema graph G_S and a keyword query Q, for each node $R \in V(G_S)$ and each subset $K' \subseteq Q$, there exists a node in G_X denoted $R\{K'\}$. For each edge $(R_1, R_2) \in E(G_S)$, and two subsets $K_1 \subseteq Q$ and $K_2 \subseteq Q$, there exists an edge $(R_1\{K_1\}, R_2\{K_2\})$ in G_X. G_X is conceptually constructed when generating CNs.

The algorithm in *S-KWS* [Markowetz et al., 2007], called InitCNGen, assigns a unique identifier to every node in G_X, and it generates all CNs by iteratively adding more nodes to a temporary result in a pre-order fashion. It does not need to check duplications using tree isomorphism for those CNs where no node, $R_i\{K'\}$, appears more than once, and it can stop enumeration of CNs from a CN, C_i, if C_i can be pruned because any CN C_j $(\supset C_i)$ must also be pruned. The general algorithm InitCNGen is shown in Algorithm 2 and the procedure CNGen is shown in Algorithm 3.

Algorithm 2 InitCNGen (Expanded Schema Graph G_X)

1: $\mathcal{C} \leftarrow \emptyset$
2: **for all** nodes $R_i \in V(G_X)$ that contain the first keyword k_1 ordered by node-id **do**
3: $\quad \mathcal{C} = \mathcal{C} \bigcup \text{CNGen}(R_i, G_X)$
4: $\quad$ remove R_i from G_X
5: **return** $\mathcal{C}$

InitCNGen makes use of Property-5, and it divides the whole CN space into several subspaces. CNs in different subspaces have different roots (start-expansion points), and CNs in the same subspace have the same root. The algorithm to generate CNs of the same root $R_i \in V(G_X)$ is shown in Algorithm 3 and will be discussed later. After processing R_i, the whole space can be divided into two subspaces as discussed in Property-5 by simply removing R_i from G_X (line 4), and

---

Algorithm 3 CNGen (Expanded Schema Node R_i, Expanded Schema G_X)

1: $\mathcal{Q} \leftarrow \emptyset; \mathcal{C} \leftarrow \emptyset$
2: Tree $C_{first} \leftarrow$ a tree of a single node R_i
3: $\mathcal{Q}.enqueue(C_{first})$
4: **while** $\mathcal{Q} \neq \emptyset$ **do**
5:      Tree $C' \leftarrow \mathcal{Q}.dequeue()$
6:      **for all** $R \in V(G_X)$ **do**
7:          **for all** $R' \in V(C')$ and $((R, R') \in E(G_X)$ or $(R', R) \in E(G_X))$ **do**
8:              **if** R can be legally added to R' **then**
9:                  Tree $C \leftarrow$ a tree by adding R as a child of R'
10:                 **if** C is a CN **then**
11:                     $\mathcal{C} = \mathcal{C} \bigcup \{C\};$ **continue**
12:                 **if** C has the potential of becoming a CN **then**
13:                     $\mathcal{Q}.enqueue(C)$
14: **return** $\mathcal{C}$

---

the unprocessed subspaces can be further divided according to the current unremoved nodes/edges in G_X. The root of trees in each subspace must contain the first keyword k_1 because each *MTJNT* will have a node that contain k_1, and it can always find a way to organize nodes in each *MTJNT* such that the node that contains k_1 is the root.

CNGen first initializes a queue $\mathcal{Q}$ and inserts a simple tree with only one node R_i into $\mathcal{Q}$ (line 1-3). It then iteratively expands a partial tree in $\mathcal{Q}$ by iteratively adding one node until $\mathcal{Q}$ becomes empty (line 4-13). At each iteration, a partial tree C' is removed from $\mathcal{Q}$ to be expanded (line 5). For every node R in G_X and R' in the partial tree C', the algorithm tests whether R can be legally added as a child of R'. Here, "legally" means

- R' must be in the rightmost root-to-leaf path in the partial tree C' (according to Property-1).

- For any node in C' that is not on the rightmost path of C', its immediate subtrees must be lexicographically ordered (according to Property-3).

- If a partial tree contains all the keywords, all the immediate subtrees for each node must be lexicographically ordered (according to Property-3), and if the root node has more than one occurrences in C', the ordered string code (canonical code) generated by the root must be the smallest among all the occurrences (according to Property-4).

If R can be legally added, then the algorithm adds R as a child of R' and forms a new tree C (line 8-9). If C itself is a CN, it outputs the tree. Otherwise, if C has the potential of becoming a CN, C will be added into $\mathcal{Q}$ for further expansion. Note that a partial tree C has the potential to become a CN if it satisfies two conditions:

- The size of $\mathcal{Q}$ must be smaller than the size control parameter Tmax.

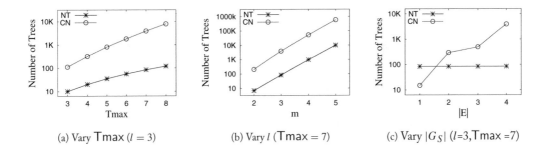

(a) Vary Tmax ($l = 3$)   (b) Vary l (Tmax $= 7$)   (c) Vary $|G_S|$ ($l=3$,Tmax $=7$)

Figure 2.6: *CN/NT* numbers on the *DBLP* Database

- Every leaf node contains a unique keyword if it is not on the rightmost root-to-leaf path in *C* (according to Property-2).

InitCNGen algorithm completely avoids the following three types of duplicates of *CN*s to be generated, comparing to the algorithm in *DISCOVER* [Hristidis and Papakonstantinou, 2002]. Isomorphic duplicates between *CN*s generated from different roots are eliminated by removing the root node from the expanded schema graph each time after calling CNGen. Duplicates that are generated from the same root following different insertion order for the remaining nodes are eliminated by the second condition in the legal node testing (line 8). The third type of duplicates occurs when the same node appears more than once in a *CN*. These types of duplicates can also be avoided by checking the third condition of the legal node testing (line 8). Avoiding the last two types of duplicates ensures that no isomorphic duplicates occur for *CN*s generated from the same root. Thus, InitCNGen generates a complete and duplication-free set of *CN*s.

The approach to generate all *CN*s in *S-KWS* [Markowetz et al., 2007] is fast when l, Tmax, and $|G_S|$ are small. The main problem with the approach is scalability: it may take hours to generate all *CN*s when $|G_S|$, Tmax, or l are large [Markowetz et al., 2007]. Note that in a real application, a schema graph can be large with a large number of relation schemas and complex foreign key references. There is also a need to be able to handle larger Tmax values. Consider a case where three authors together write a paper in the *DBLP* database with schema shown in Figure 2.1. The smallest number of tuples needed to include an *MTJNT* for such a case is Tmax $= 7$ (3 Author tuples, 3 Write tuples, and 1 Paper tuple).

Figure 2.6 shows the number of *CN*s, denoted CN, for the *DBLP* database schema (Figure 2.1). Given the entire database schema, Figure 2.6(a) shows the number of *CN*s by varying Tmax when the number of keywords is 3, and Figure 2.6(b) shows the number of *CN*s by varying the number of keywords, l when Tmax $= 7$. Figure 2.6(c) shows the number of *CN*s by varying the complexity of the schema graph (Figure 2.1). Here, the 4 points on x-axis represent four cases: Case-1 (Author and Write with foreign key reference between the two relation schemas), Case-2 (Case-1 plus Paper

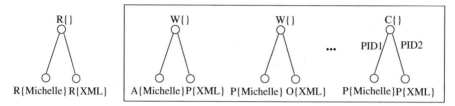

Figure 2.7: An *NT* that represents many *CN*s

with foreign key reference between `Write` and `Paper`), Case-3 (Case-2 plus `Cite` with one of the two foreign key references between `Paper` and `Cite`), and Case-4 (Case-2 with both foreign key references between `Paper` and `Cite`). For the simple database schema with 4 relation schemas and 4 foreign key references, the number of *CN*s increases exponentially. For example, when $l = 5$ and $\mathsf{Tmax} = 7$, the number of *CN*s is about 500,000.

In order to significantly reduce the computational cost to generate all *CN*s, a new fast template-based approach can be used. In brief, we can first generate all *CN* templates (candidate network templates or simply network templates), denoted *NT*, and then generate all *CN*s based on all *NT*s generated. In other words, we do not generate all *CN*s directly like InitCNGen in *S-KWS* [Markowetz et al., 2007]. The cost saving of this approach is high. Recall that given an *l*-keyword query against a database schema G_S, there are $2^l \cdot |V(G_S)|$ nodes (relations), and, accordingly, there are $2^{2l} \cdot |E(G_S)|$ edges in total in the extended graph G_X. There are two major components that contribute to the high overhead of InitCNGen.

- **(Cost-1)** The number of nodes in G_X that contain a certain selected keyword k is $|V(G_S)| \cdot 2^{l-1}$ (line 1). InitCNGen treats each of these nodes, n_i, as the root of a *CN* cluster and calls CNGen to find all valid *CN*s starting from n_i.

- **(Cost-2)** The CNGen algorithm expands a partial *CN* edge-by-edge based on G_X at every iteration and searches all *CN*s whose size is $\leq \mathsf{Tmax}$. Note that in the expanded graph G_X, a node is connected to/from a large number of nodes. CNGen needs to expand **all** possible edges that are connected to/from every node (refer to line 8 in CNGen).

In order to reduce the two costs, in the template based approach, a template, *NT*, is a special *CN* where every node, $R\{K'\}$, in *NT* is a variable that represents any sub-relation, $R_i\{K'\}$. Note that a variable represents $|V(G_S)|$ sub-relations. A *NT* represents a set of *CN*s. An example is shown in Figure 2.7. The leftmost is a *NT*, $R\{\text{Michelle}\} \bowtie R\{\} \bowtie R\{\text{XML}\}$, shown as a tree rooted at $R\{\}$. There are many *CN*s that match the *NT* as shown in Figure 2.7. For example, $A\{\text{Michael}\} \bowtie W\{\} \bowtie P\{\text{XML}\}$ and $P\{\text{Michael}\} \bowtie C\{\} \bowtie P\{\text{XML}\}$ match the *NT*. The number of *NT*s is much smaller than the number of *CN*s, as indicated by NT in Figure 2.6(a) (b) and (c). When $l = 5$ and $\mathsf{Tmax} = 7$, there are 500,000 *CN*s but only less than 10,000 *NT*s.

Algorithm 4 InitNTGen(The schema graph G_S)

1: $C \leftarrow \emptyset$

2: $\mathcal{T} \leftarrow$ InitCNGen(G_1)

3: **for all** $T \in \mathcal{T}$ **do**

4:     **for all** nodes $R_i \in V(G_S)$ **do**

5:         $C \leftarrow C \cup$ NT2CN(T, R_i, G_S)

6: **return** C

7: **Procedure** NT2CN(Tree T, Relation R_i, The schema graph G_S)

8: $r \leftarrow root(T)$

9: $C' \leftarrow \{R_i\{K'\}\}$ if r is with a subset of keywords K'

10: **for all** immediate child of r, s, in T **do**

11:     $C'' \leftarrow \emptyset$

12:     **for all** $\{R_j | (R_i, R_j) \in E(G_S) \vee (R_j, R_i) \in E(G_S)\}$ **do**

13:         $C'' \leftarrow C'' \bigcup$ NT2CN($subtree(s), R_j, G_S$)

14:     $C' \leftarrow C' \times C''$

15:     prune C' using (Rule-3)

16: **return** C'

NT Generation: We can generate all NTs using a slightly modified InitCNGen algorithm against a special schema graph $G_1(V, E)$, where V only consists of a single node, R, and there is only one edge between R and itself. The relation schema R is with an attribute which is both primary key and foreign key referencing to its primary key.[2] With the special graph schema G_1, both costs (Cost-1 and Cost-2) can be significantly reduced in generating all NTs. InitCNGen needs to call CNGen $|V(G_s)| \cdot 2^{l-1}$ times. The template-based approach only calls it 2^{l-1} times. For Cost-2, since there is only one edge in the special schema graph G_1, we only need to check one possible edge when expanding.

CN Generation: For every generated NT, we can further generate all CNs that match it using a dynamic programming algorithm, which can be done fast. We will discuss it below in detail. It is important to note that the complexity of the schema graph, $|G_S|$, does not have a significant impact on the computational cost to generate CNs using the template-based approach.

The template-based algorithm to generate CNs is illustrated in Algorithm 4. It first initializes the set of CNs, C, to be empty (line 1). Then, it generates all NTs, $\mathcal{T}$, by calling the InitCNGen algorithm [Markowetz et al., 2007] with a special one node and one edge graph, G_1. For every network template $T \in \mathcal{T}$, it attempts to generate all CNs that match T by enumerating all the relations, R_i, in the schema graph, G_S. It calls NT2CN for every template T and every relation (line 3-5). The set of CNs, C, will be returned at the end.

---

[2]Note that in *S-KWS*, InitCNGen can handle a database schema with self-loops. By labeling an edge with a foreign key, InitCNGen can handle parallel edges (multiple foreign key references) between two nodes in G_S.

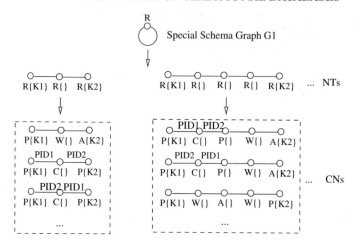

Figure 2.8: Generate all *CN*s using *NT*s

We can formalize every template T as a rooted tree, and we can, therefore obtain a set of *CN*s as a set of rooted trees that match some template. Here, the root of a *NT*/*CN* tree is node r in the tree such that the maximum path from r to its leaf nodes is minimized. Such a root can be quickly identified as the center of the tree by removing all the leaf nodes in every iteration repeatedly until there is either a single node (one-center) or two nodes with a single edge left (two-centers). If a tree has one-center, the center is the root, which is unique. If a tree has two-centers, we select the node that has a smaller node-identifier as the root. The height of any *NT* is at most Tmax/2 + 1.

The procedure NT2CN (lines 7-16, Algorithm 4) assigns real labels such as relation names to a *NT* and, therefore, generates all *CN*s that match a given *NT*. In NT2CN, for a given *NT*, T, and a relation R_i, it first identifies the root of T, as r (line 8). Because a node in a *NT* is associated with a set of keywords K', $R_i\{K'\}$ is a possible match to match the root node r, and $R_i\{K'\}$ is assigned to C' (line 9). Next, it considers the child nodes of r, indicated as s, in T, and it repeats the same attempts to label s with a relation name by recursively calling NT2CN (lines 10-13). For any return results, C'', of the union for recursive calls (line 13), it conducts $C' \times C''$ (line 14) followed by pruning using the same (Rule-3) to prune any partial invalid *CN*s.

The whole process of InitNTGen is illustrated in Figure 2.8. Starting from a special graph G_1, all *NT*s are generated. For each *NT*, a set of *CN*s are generated that match the *NT*. The new InitNTGen algorithm generates a complete and duplication-free set of *CN*s.

The complexity of schema graph, G_S, does not have a significant impact on the efficiency of the InitNTGen algorithm. The efficiency of InitNTGen is determined by two factors, Tmax and l. This is due to the fact that InitNTGen does not generate a large expanded graph G_X to work with. It instead starts from a smaller graph G_1. We discuss the issues related to time complexity below.

The number of CNs is exponentially large with respect to three factors, namely, the number of keywords in query l, the size control parameter Tmax, and schema complexity $|G_S|$. The time/space complexity for both InitCNGen and the new InitNTGen are difficult to derive because the number of CNs is not easy to determine. Below, we explain why InitNTGen is much better than InitCNGen by showing that InitNTGen can avoid checking an exponential number of partial results, which is costly. First, we generate all NTs using InitCNGen upon a special schema graph G_1 with only 1 node. The time complexity to generate all NTs on G_1 is much smaller than the time complexity to generate all CNs on the original graph G_S using InitCNGen because the former does not need to consider the exponential number of partial CNs with respect to the schema complexity $|G_S|$. The main cost saving is on evaluating all NTs using NT2CN. For NT2CN, although we can not avoid generating all CNs that are in the final result, we avoid generating a large number of partial CNs that can not potentially result in a valid CN or will result in redundant CNs. Recall that CNGen (Algorithm 3), before expanding the partial CN by one node u, needs to check whether u can be legally added to the partial CN. "Legally added" means that node u can be inserted only to the rightmost root-to-leaf path, and it must be no smaller than any of its siblings. Furthermore, if after adding u, the partial tree contains all the keywords, all the immediate subtrees for each node must be lexicographically ordered. Therefore, it needs $O(\text{Tmax}^2)$ time to check whether u can be legally added. There are $O(2^l \cdot |V(G_S)|)$ nodes in total to be checked, and it is highly possible that only few of nodes can be legally added. In the worst case, it needs $O(2^l \cdot |V(G_S)| \cdot \text{Tmax}^2)$ time to check whether it can add one node to a partial CN. Reducing this cost can be crucial to the total time complexity of generating all CNs. In the template based algorithm, in generating all NTs, for each partial NT, there may be an exponential number of partial CNs that can match it. We use the same operations to check, but we only need to check such a partial NT once, because in NT2CN, all CNs generated are complete and duplication-free, and no such checking operations are needed. As a result, we can avoid an exponential number of such costly checking operations with respect to l, $|G_S|$ and Tmax.

2.3   CANDIDATE NETWORK EVALUATION

After generating all candidate networks (CNs) in the first phase, the second phase is to evaluate all candidate networks in order to get the final results. There are two situations:

- **Getting all *MTJNT*s in a relational database:** As studied in *DBXplorer* [Agrawal et al., 2002], *DISCOVER* [Hristidis and Papakonstantinou, 2002], *S-KWS* [Markowetz et al., 2007], *KDynamic* [Qin et al., 2009c] and *KRDBMS* [Qin et al., 2009a], all *MTJNT*s are evaluated upon the set of CNs generated by specifying a proper execution plan.

- **Getting top-k *MTJNT*s in a relational database:** As studied in *DISCOVER-II* [Hristidis et al., 2003a] and *SPARK* [Luo et al., 2007], it is ineffective to present users a huge number of *MTJNT*s generated. In this situation, only top-k *MTJNT*s are presented according to Definition 2.6.

Algorithm 5 Discover-CNEval(The set of *CNs* *C*)

1:   evaluate all subexpressions of size 1
2:   **while** not all *CN*s in *C* are evaluated **do**
3:     $\mathcal{E} = \{e | e$ is a subexpression in C and $e = e_1 \bowtie e_2$ and e_1, e_2 are evaluated subexpressions$\}$
4:     e = subexpression in $\mathcal{E}$ that has the highest score
5:     evaluate e

2.3.1   GETTING ALL *MTJNT*S IN A RELATIONAL DATABASE

In RDBMS, the problem of evaluating all *CN*s in order to get all *MTJNT*s is a multi-query optimization problem. There are two main issues: (1) how to share common subexpressions among *CN*s generated in order to reduce computational cost when evaluating. We call the graph the *execute graph* that consists of all *CN*s and is obtained as a result of sharing computational cost, (2) how to find a proper join order on the *execute graph* to fast evaluate all *CN*s. For a keyword query, the number of *CN*s generated can be very large. Given a large number of joins, it is extremely difficult to obtain an optimal query processing plan because one best plan for a *CN* may slow down others if its subtrees are shared by other *CN*s. As studied in *DISCOVER* [Hristidis and Papakonstantinou, 2002], finding the optimal execution plan is an NP-complete problem.

    In *DISCOVER* [Hristidis and Papakonstantinou, 2002], an algorithm is proposed to evaluate all *CN*s together using a greedy algorithm based on the following observations: (1) subexpressions (intermediate *CN*s) that are shared by most *CN*s should be evaluated first; and (2) subexpressions that may generate the smallest number of results should be evaluated first. We assume the statistics of the relations in the database are known and the size of results for any subexpression e is denoted as $estimite(e)$ and the number of *CN*s that share the subexpression e is denoted as $frequency(e)$. For any subexpression e, let $score(e) = \frac{frequency(e)^a}{\log^b(estimite(e))}$, where a and b are constants. The *DISCOVER* algorithm to evaluate all *CN*s is shown in Algorithm 5.

    The algorithm first evaluates all subexpressions of size 1 such as W, A, $A\{Michelle\}$ and so on in the running example. Then it iteratively evaluates a subexpression with the highest score value as discussed above until all *CN*s are evaluated. The newly evaluated subexpression must be a join with two subexpressions already evaluated.

    Sharing computational cost for subexpressions is also discussed in many other papers. In *S-KWS*, in order to share the computational cost of evaluating all *CN*s, Markowetz et al. construct an operator mesh. In a mesh, there are $n \cdot 2^{l-1}$ clusters, where n is the number of relations in the schema graph G_S and l is the number of keywords. A cluster consists of a set of operator trees (left-deep trees) that share common expressions. More precisely, each cluster is the compressed representation of the set of *CN*s generated each time line 3 of Algorithm 2 is invoked, where the left part of each left-deep tree includes the root that contains the first shared keyword k_1. A left-deep tree is shown in Figure 2.9(b), where the leaf nodes are the projections, and the non-leaf nodes are joins. The output of a node (project/join) can be shared by other left-deep trees as input. Two partial clusters in a mesh

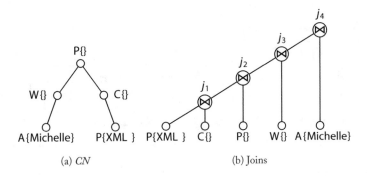

(a) *CN*   (b) Joins

Figure 2.9: A *CN* and a Left Deep Tree

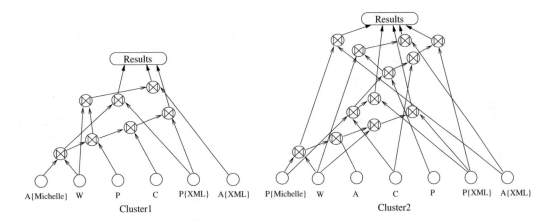

Figure 2.10: Two Partial Clusters in a Mesh

are shown in Figure 2.10, the first cluster includes all *CN*s rooted at $A\{\text{Michelle}\}$ and the second cluster includes all *CN*s rooted at $P\{\text{Michelle}\}$ by removing $A\{\text{Michelle}\}$ from the expanded schema graph G_X. When evaluating all *CN*s in a mesh, a projected relation with the smallest number of tuples is selected to start and to join.

In *KDynamic* [Qin et al., 2009c], a new structure called an $\mathcal{L}$-Lattice is introduced to share computational cost among *CN*s. Given a set of *CN*s, $\mathcal{C}$, we define the root of each *CN* to be the node r such that the maximum path from r to all leaf nodes of the *CN* is minimized. The $\mathcal{L}$-Lattice is constructed as follows: when a new rooted *CN*, C_i, is inserted to $\mathcal{L}$, we generate the canonical codes for all its rooted subtrees of the rooted *CN* tree, C_i, including C_i itself. Recall that a canonical code is a string. Two trees, C_i and C_j, are identical if and only if their canonical codes are identical. We index all subtrees in $\mathcal{L}$ using their canonical codes over $\mathcal{L}$, while constructing $\mathcal{L}$. For a given rooted

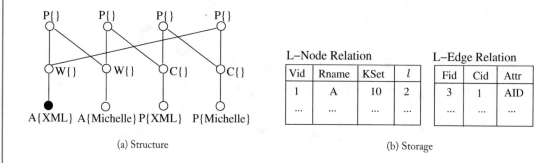

(a) Structure

L–Node Relation

Vid	Rname	KSet	l
1	A	10	2
...	...	...	...

L–Edge Relation

Fid	Cid	Attr
3	1	AID
...	...	...

(b) Storage

Figure 2.11: $\mathcal{L}$-Lattice

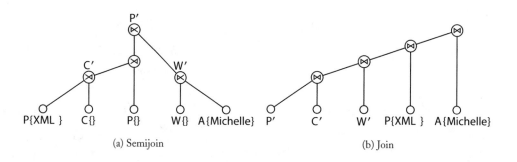

(a) Semijoin

(b) Join

Figure 2.12: Join vs Semijoin/Join

CN C_i, we attempt to find the largest subtrees in $\mathcal{L}$ that C_i can share with using the index, and we link to the roots of such subtrees. Figure 2.11(a) illustrates a partial lattice. The entire lattice, $\mathcal{L}$, is maintained in two relations: L-Node relation and L-Edge relation (Figure 2.11(b)). Let a bit-string represent a set of keywords, $\{k_1, k_2, \cdots, k_l\}$. The L-Node relation maintains, for any node in $\mathcal{L}$, a unique Vid in $\mathcal{L}$, the corresponding relation name (Rname) that appears in the given database schema, G_S, a bit-string (KSet) that indicates the keywords associated with the node in $\mathcal{L}$, and the size of the bit-string (l). The L-Edge relation maintains the parent/child relations among all the nodes in $\mathcal{L}$ with its parent Vid and child Vid (Fid/Cid) plus its join attribute, Attr, (either primary key or foreign key). The two relations can be maintained in memory or on disk. Several indexes are built on the relations to quickly search for given nodes in $\mathcal{L}$.

There are three main differences between the two *execute graphs*: the Mesh and the $\mathcal{L}$-Lattice. (1) The maximum depth of a Mesh is Tmax $- 1$ and the maximum depth of an $\mathcal{L}$-Lattice is $\lfloor$Tmax$/2 + 1\rfloor$. (2) In a mesh, only the left part of two *CN*s can be shared (except for the leaf nodes), while in an $\mathcal{L}$-Lattice multiple parts of two *CN*s can be shared. (3) The number of leaf nodes in a mesh is $O((|V(G_S)| \cdot 2^l)^2)$ because there are $O(|V(G_S)| \cdot 2^l)$ clusters in a mesh and each cluster may contain $O(|V(G_S)| \cdot 2^l)$ leaf nodes. The number of leaf nodes in an $\mathcal{L}$-Lattice is $O(2^l)$.

After sharing computational cost using either the Mesh or the $\mathcal{L}$-Lattice, all *CN*s are evaluated using joins in *DISCOVER* or *S-KWS*. A join plan is shown in Figure 2.9(b) to process the *CN* in Figure 2.9(a) using 5 projects and 4 joins. The resulting relation, the output of the join (j_4), is a temporal relation with 5 TIDs from the 5 projected relations, where a resulting tuple represents an *MTJNT*. The rightmost two connected trees in Figure 2.3 are the two results of the operator tree Figure 2.9(b), $(p_2, c_5, p_4, w_5, a_3)$ and $(p_3, c_4, p_4, w_5, a_3)$.

In *KRDBMS* [Qin et al., 2009a], the authors observe that evaluating all *CN*s using only joins may always generate a large number of temporal tuples. They propose to use semijoin/join sequences to compute a *CN*. A semijoin between R and S is defined in Eq. 2.18, which is to project (Π) the tuples from R that can possibly join at least a tuple in S.

$$R \ltimes S = \Pi_R(R \bowtie S) \tag{2.18}$$

Based on semijoin, a join $R \bowtie S$ can be supported by a semijoin and a join as given in Eq. 2.19.

$$R \bowtie S = (R \ltimes S) \bowtie S \tag{2.19}$$

Recall that semijoin/joins were proposed to join relations in a distributed RDBMS, in order to reduce high communication cost at the expense of I/O cost and CPU cost. But, there is no communication in a centralized RDBMS. In other words, there is no obvious reason to use $(R \ltimes S) \bowtie S$ to process a single join $R \bowtie S$ since the former needs to access the same relation S twice. Below, we address the significant cost saving of semijoin/joins over joins when the number of joins is large, in a centralized RDBMS.

When evaluating all *CN*s, the temporal tuples generated can be very large, and the majority of the generated temporal tuples do not appear in any *MTJNT*s. When evaluating all *CN*s using the semijoin/join based strategy, computing $R \bowtie (S \bowtie T)$ is done as $S' \leftarrow S \ltimes T$, $R' \leftarrow R \ltimes S'$, with semijoins, in the reduction phase, followed by $T \bowtie (S' \bowtie R')$ in the join phase. For the *CN* given in Figure 2.9(a), in the reduction phase (Figure 2.12(a)), $C' \leftarrow C\{\} \ltimes P\{\text{XML}\}$, $W' \leftarrow W\{\} \ltimes A\{\text{Michelle}\}$, $P'' \leftarrow P\{\} \ltimes C'$, and $P' \leftarrow P'' \ltimes W'$, and in the join phase (Figure 2.12(b)), $P' \bowtie C'$ is joined first because P' is fully reduced, such that every tuple in P' must appear at an *MTJNT*. The join order is shown in Figure 2.12(b).

Figure 2.13 shows the number of temporal tuples generated using a real database *DBLP* on IBM *DB2*. The five 3-keyword queries with different keyword selectivity (the probability that a tuple contains a keyword in *DBLP*) were randomly selected with Tmax = 5. The number of generated temporal tuples are shown in Figure 2.13(a). The number of tuples generated by the semijoin/join approach is significantly less than that by the join approach. In a similar fashion, the number of temporal tuples generated by the semijoin/join approach is significantly less than that generated by the join approach when Tmax increases (Figure 2.13(b)) for a 3-keyword query.

When processing a large number of joins for keyword search on RDBMSs, it is the best practice to process a large number of small joins in order to avoid intermediate join results becoming very

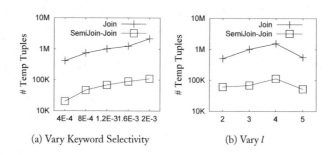

(a) Vary Keyword Selectivity   (b) Vary l

Figure 2.13: # of Temporal Tuples (Default $\mathsf{Tmax} = 5, l = 3$)

large and dominative if it is difficult to find an optimal query processing plan or the cost of finding an optimal query processing plan is high.

Besides evaluating all *CN*s in a static environment, *S-KWS* and *KDynamic* focus on monitoring all *MTJNT*s in a relational data stream where tuples can be inserted/deleted frequently. In this situation, it is necessary to find new *MTJNT*s or expire *MTJNT*s in order to monitor events that are implicitly interrelated over an open-ended relational data stream for a user-given l-keyword query. More precisely, it reports new *MTJNT*s when new tuples are inserted, and, in addition, reports the *MTJNT*s that become invalid when tuples are deleted. A sliding window (time interval), W, is specified. A tuple, t, has a lifespan from its insertion into the window at time $t.start$ to $W + t.start - 1$, if t is not deleted before then. Two tuples can be joined if their lifespans overlap.

S-KWS processes a keyword query in a relational data stream using the mesh as introduced above. The authors observe that in a data stream environment some joins need to be processed when there are incoming new tuples from its inputs but not all joins need to be processed all the time, and, therefore, they propose a demand-driven operator execution. A join operator has two inputs and is associated with an output buffer. The output buffer of a join operator becomes input to many other join operators that share the join operator (as indicated in the mesh). A tuple that is newly output by a join operator in its output buffer will be a new arrival input to those joins that share the join operator. A join operator will be in a running state if it has newly arrived tuples from both inputs. A join operator will be in a sleeping state if either it has no new arriving tuples from the inputs or all the join operators that share it are currently sleeping. The demand-driven operator execution noticeably reduces the query processing cost.

KDynamic processes a keyword query in a relational data stream using the $\mathcal{L}$-Lattice. Although *S-KWS* can significantly reduce the computational cost, the scalability issues is also a problem especially when Tmax, $|G_S|$, l, W or the stream speed is high. This is because a large number of intermediate tuples that are computed by many join operators in the mesh with high processing cost will eventually not be output. *S-KWS* cannot avoid computing such a large number of unnecessary intermediate tuples because it is unknown whether an intermediate tuple will appear in an *MTJNT* beforehand. The probability of generating a large number of unnecessary intermediate results increases when either the size of sliding window, W, is large, or new data arrive at high speed. It is challenging to reduce the processing cost by reducing the number of intermediate results.

In *KDynamic*, an algorithm CNEvalDynamic is proposed, which works as follows. We can maintain $|V(G_S)|$ relations in total to process an l-keyword query $Q = \{k_1, k_2, \cdots, k_l\}$, due to the lattice structure that is used. A node, v, in lattice $\mathcal{L}$ is uniquely identified with a node id. The node v represents a sub-relation $R_i\{K'\}$. By utilizing the unique node id, it is easy to maintain all the 2^l sub-relations for a relation R_i together. Let us denote such a relation as $\mathbf{R}_i$. The schema of $\mathbf{R}_i$ is the same as R_i plus an additional attribute (Vid) to keep the node id in $\mathcal{L}$. When we need to obtain a sub-relation $R_i\{K'\}$ for $K' \subseteq Q$ associated with a node, v, in the lattice, we use the node id to select and project $R_i\{K'\}$ from $\mathbf{R}_i$. Therefore, a relation $R_i\{K'\}$ can be possibly virtually maintained. Below, we use $\mathbf{R}_i\{K'\}$ to denote such a sub-relation. It is fast to obtain $\mathbf{R}_i\{K'\}$ if an index is built on the additional attribute Vid on relation $\mathbf{R}_i$.

CNEvalDynamic is outlined in Algorithm 6. When a new update operator, $op(t, R_i)$, arrives, it processes it in lines 3-9 if the operation is an insertion or in lines 11-14 if it is a deletion. The procedure EvalPath joins all the needed tuples in a top-down fashion. EvalPath is implemented similar to the semijoin-join based static evaluation as discussed above using an additional *path*, which records where the join sequence comes from to reduce join cost. The two procedures, namely insert and delete, maintain a list of tuples for each node in the lattice using only selections (lines 17-18, lines 26-27, and lines 34-35). The selected tuples can join at least one tuple from each list of its child nodes in the lattice. If the list of one node in the lattice is changed, it will trigger the father nodes to change their lists accordingly (lines 24-27 and lines 32-35). If the root node is changed, this means the results should be updated. At this time, we use joins to report the updated *MTJNTs*. When we join, all the tuples that participate in joins will contribute to the results. In this way, we can achieve full reduction when joining.

As the number of results itself can be exponentially large, we analyze the extra cost for the algorithms to evaluate all *CN*s. The extra cost is defined to be the number of tuples generated by the algorithm minus the number of tuples in the result. Suppose the number of tuples in every relation is n. Given a *CN* with size t, the extra cost for the algorithm using the left deep tree proposed in *S-KWS* to evaluate the *CN* is $O(n^{t-1})$, and the extra cost for the CNEvalDynamic algorithm to evaluate the *CN* is $O(n \cdot t)$.

Finally, we discuss how to implement the event-driven evaluation. As shown in Figure 2.14, there are multiple nodes labeled with identical $R_i\{K'\}$. For example, $W\{\}$ appears in two different nodes in the lattice. For each $R_i\{K'\}$, we maintain 3 lists named Rlist (Ready), Wlist (Wait) and Slist (Suspend). The three lists together contain all the node ids in the lattice. A node in the lattice $\mathcal{L}$ labeled $R_i\{K'\}$ can only appear in one of the three lists for $R_i\{K'\}$. A node v in $\mathcal{L}$ appears in Wlist, if the sub-relations represented by all child nodes of v in $\mathcal{L}$ are non-empty, but the sub-relation represented by v is empty. A node v in $\mathcal{L}$ appears in Rlist, if the sub-relations represented by all child nodes of v in $\mathcal{L}$ are non-empty, and the sub-relation represented by v itself

---

Algorithm 6 CNEvalDynamic($\mathcal{L}$, Q, Σ)

Input: An l-keyword query Q, a lattice $\mathcal{L}$, and a set of *MTJNT*s denoted Σ

1: **while** a new update $op(t, R_i)$ arrives **do**
2:   let K' be the set of all keywords appearing in tuple t
3:   **if** op is to insert t into relation R_i **then**
4:     $\Delta \leftarrow \emptyset$
5:     **for** each v in $\mathcal{L}$ labeled $R_i\{K'\}$ **do**
6:       $path \leftarrow \emptyset$
7:       insert(v, t, Δ)
8:       report new *MTJNT*s in Δ
9:       $\Sigma \leftarrow \Sigma \cup \Delta$
10:   **else if** op is to delete t from R_i **then**
11:     **for** each v in $\mathcal{L}$ labeled $R_i\{K'\}$ **do**
12:       **if** $t \in \mathbf{R}_i\{K'\}$ **then**
13:         delete(v, t)
14:         delete *MTJNT*s in Σ that contain t, and report such deletions

15: **Procedure** insert(v, t, Δ)
16: let the label of node v be $R_i\{K'\}$
17: **if** $t \notin \mathbf{R}_i\{K'\}$ and t can join at least one tuple in every relation represented by all v's children in $\mathcal{L}$ **then**
18:   insert tuple t into the sub-relation $\mathbf{R}_i\{K'\}$
19: **if** $t \in \mathbf{R}_i\{K'\}$ **then**
20:   push (v, t) to $path$
21:   **if** v is a root node in $\mathcal{L}$ **then**
22:     $\Delta \leftarrow \Delta \cup$ EvalPath$(v, t, path)$
23:   **else**
24:     **for** each father node of v, u, in $\mathcal{L}$ **do**
25:       let the label of node u be $R_j\{K''\}$
26:       **for** each tuple t' in $\pi_{K''}(r(R_j))$ that can join t **do**
27:         insert(u, t', Δ)
28:   pop (v, t) from $path$

29: **Procedure** delete(v, t)
30: let the label of node v be $R_i\{K'\}$
31: delete tuple t from the sub-relation $\mathbf{R}_i\{K'\}$
32: **for** each father node of v, u in $\mathcal{L}$ **do**
33:   let the label of node u be $R_j\{K''\}$
34:   **for** each tuple t' in $\mathbf{R}_j\{K''\}$ that can join t only **do**
35:     delete(u, t')

---

is non-empty too. Otherwise, v appears in `Slist`. When a new tuple t of relation R_i with keyword set K' is inserted, we only insert it into all relations in the nodes v, in $\mathcal{L}$, on `Rlist` and `Wlist` specified for $R_i\{K'\}$. Each insertion may notify some father nodes of v to move from `Wlist` or `Slist` to `Rlist`. Node v may also be moved from `Wlist` to `Rlist`. When a tuple t of relation R_i with keyword set K' is about to be deleted, we only remove it from all relations associated with node

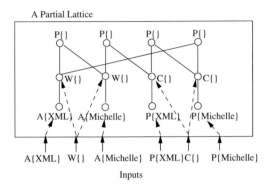

Figure 2.14: Lattice and Its Inputs from a Stream

v, in $\mathcal{L}$, on `Rlist` specified for $R_i\{K\}$. Each deletion may notify some father nodes of v to be moved from `Rlist` or `Wlist` to `Slist`, and v may also be moved from `Rlist` to `Wlist`.

2.3.2   GETTING TOP-k *MTJNT*S IN A RELATIONAL DATABASE

We have discussed several effective ranking strategies in Section 2.1. In this section, we discuss how to answer the top-k keyword queries efficiently. A naive approach is to first generate all *MTJNT*s using the algorithms proposed in Section 2.3.1, and then calculate the score for each *MTJNT*, and finally output the top-k *MTJNT*s with the highest scores. In *DISCOVER-II* [Hristidis et al., 2003a] and *SPARK* [Luo et al., 2007], several algorithms are proposed to get top-k *MTJNT*s efficiently. The aim of all the algorithms is to find a proper order of generating *MTJNT*s in order to stop early before all *MTJNT*s are generated.

In *DISCOVER-II*, three algorithms are proposed to get top-k *MTJNT*s, namely, the Sparse algorithm, the Single-Pipelined algorithm, and the Global-Pipelined algorithm. All algorithms are based on the attribute level ranking function given in Eq. 2.1. Given a keyword query Q, for any tuple t, let the tuple score be $score(t, Q) = \sum_{a \in t} score(a, Q)$ where $score(a, Q)$ is the score for attribute a of t as defined in Eq. 2.2. The score function in Eq. 2.1 has the property of *tuple monotonicity*, defined as follows. For any two *MTJNT*s $T = t_1 \bowtie t_2 \bowtie ... \bowtie t_l$ and $T' = t'_1 \bowtie t'_2 \bowtie ... \bowtie t'_l$ generated from the same *CN* C, if for any $1 \le i \le l$, $score(t_i, Q) \le score(t'_i, Q)$, then we have $score(T, Q) \le score(T', Q)$.

For a keyword query Q, given a *CN* C, let the set of keyword relations that contain at least one keyword in C be $C.M = \{M_1, M_2, ..., M_s\}$. Suppose tuples in each M_i ($1 \le i \le s$) are sorted in non-increasing order of their scores. Let $M_i.t_j$ be the j-th tuple in M_i. In each M_i, we use $M_i.cur$ to denote the current tuple such that the tuples before the position of the tuple are all accessed, and we use $M_i.cur \leftarrow M_i.cur + 1$ to move $M_i.cur$ to the next position. We use $eval(t_1, t_2, ..., t_s)$ (where t_i is a tuple and $t_i \in M_i$) to denote the *MTJNT*s of C by fixing M_i to be t_i. It can be done by issuing an SQL statement in RDBMS. We use $\overline{score}(C, Q)$ to denote the upper bound score for all

---

Algorithm 7 Sparse (the keyword query Q, the top-k value k)

1: $topk \leftarrow \emptyset$
2: **for all** CNs C ranked in decreasing order of $\overline{score}(C, Q)$ **do**
3:      **if** $score(topk[k], Q) \geq \overline{score}(C, Q)$ **then**
4:         **break**
5:      evaluate C and update $topk$
6: **output** $topk$

$MTJNT$s in C, defined as follows:

$$\overline{score}(C, Q) = \sum_{i=1}^{s} score(M_i.t_1, Q) \tag{2.20}$$

The Sparse **Algorithm:** The Sparse algorithm avoids evaluating unnecessary CNs which can not possible generate results that are ranked top-k. The algorithm is shown in Algorithm 7. It first sorts all CNs by their upper bound value $\overline{score}(C, Q)$, then for each CN, it generates all its $MTJNT$s and uses them to update $topk$ (line 5). If the upper bound of the next CN is no larger than the k-th largest score $score(top[k], Q)$ in the $topk$ list, it can safely stop and output $topk$ (lines 3-4).

The Single-Pipelined **Algorithm:** Given a keyword query Q, the Single-Pipelined algorithm first gets the top-k $MTJNT$s for each CN, and then combines them together to get the final result. Suppose $C.M = \{M_1, M_2, ..., M_s\}$ for a given CN C, and let $\overline{score}(C.M, i)$ denote the upper bound score for any $MTJNT$s that include the unseen tuples in M_i. We have:

$$\overline{score}(C.M, i) = \sum_{1 \leq j \leq s \text{ and } j \neq i} score(M_j.t_1, Q) + score(M_i.cur + 1, Q) \tag{2.21}$$

The Single-Pipelined algorithm (Algorithm 8) works as follows. Initially, all tuples in M_i $(1 \leq i \leq s)$ are unseen except for the first one, which is used for upper bounding the other unseen tuples (lines 2-4). Then, it iteratively chooses the list M_p that maximizes the upper bound score, and it moves $M_p.cur$ to the next unseen tuple (lines 6-7). It processes $M_p.cur$ using all the seen tuples in other lists M_i $(i \neq p)$ and uses the results to update $topk$ (lines 8-9). If once the maximum possible upper bound score for all unseen tuples $max_{1 \leq i \leq s} \overline{score}(C.M, i)$ is already no larger than the k-th largest score in the $topk$ list, it can safely stop and output $topk$ (line 5).

The Global-Pipelined **Algorithm:** The Single-Pipelined algorithm introduced above considers each CN individually before combining their top-k results in order to get the final top-k results. The Global-Pipelined algorithm considers all the CNs together. It uses similar procedures as the Single-Pipelined algorithm. The only difference is that, there is only one $topk$ list, and each time, it selects a CN C_p such that $max_{1 \leq i \leq s} \overline{score}(C_p.M, i)$ is maximized before processing lines 6-9 in the Single-Pipelined algorithm. Once the upper bound value for all unseen tuples

---

Algorithm 8 Single-Pipelined (the keyword query Q, the top-k value k, the CN C)

1: $topk \leftarrow \emptyset$
2: let $C.M = \{M_1, M_2, ..., M_s\}$
3: initialize $M_i.cur \leftarrow M_i.t_1$ for $1 \leq i \leq s$
4: update $topk$ using $eval(M_1.t_1, M_2.t_1, ..., M_s.t_1)$
5: **while** $max_{1 \leq i \leq s} \overline{score}(C.M, i) > score(topk[k], Q)$ **do**
6:     suppose $\overline{score}(C.M, p) = max_{1 \leq i \leq s} \overline{score}(C.M, i)$
7:     $M_p.cur \leftarrow M_P.cur + 1$
8:     **for all** $t_1, t_2, ..., t_{p-1}, t_{p+1}, ..., t_s$ where t_i is seen and $t_i \in M_i$ for $1 \leq i \leq s$ **do**
9:         update $topk$ using $eval(t_1, t_2, ..., t_{p-1}, M_p.cur, t_{p+1}, ..., t_s)$
10: **output** $topk$

$max_{1 \leq i \leq s, C_j \in C} \overline{score}(C_j.M, i)$ is no larger than the k-th largest value in the $topk$ list, it can stop and output the global top-k results.

In *SPARK* [Luo et al., 2007], the authors study the tree level ranking function Eq. 2.11. This ranking function does not satisfy tuple monotonicity. As a result, the earlier discussed top-k algorithms that stop early (e.g., the Global-Pipelined algorithm) can not be insured to output correct top-k results. In order to handle such non-monotonic score functions, a new monotonic upper bound function is introduced. The intuition behind the upper bound function is that, if the upper bound score is already smaller than the score of a certain result, then all the upper bound scores of unseen tuples will be smaller than the score of this result due to the monotonicity of the upper bound function. The upper bound score $uscore(T, Q)$ is defined as follows:

$$uscore(T, Q) = uscore_a(T, Q) \cdot score_b(T, Q) \cdot score_c(T, Q) \tag{2.22}$$

where

$$uscore_a(T, Q) = \frac{1}{1-s} \cdot min(A(T, Q), B(T, Q))$$

$$A(T, Q) = sumidf(T, Q) \cdot (1 + \ln(1 + \ln(\sum_{t \in T} wantf(t, T, Q))))$$

$$B(T, Q) = sumidf(T, Q) \cdot \sum_{t \in T} watf(t, T, Q)$$

$$sumidf(T, Q) = \sum_{w \in T \cap Q} idf(T, w)$$

$$wantf(t, T, Q) = \frac{\sum_{w \in t \cap Q} tf(t, w) \cdot idf(T, w)}{sumidf(T, Q)}$$

$score_b(T, Q)$ and $score_c(T, Q)$ can be determined given the CN of T. We have the follow Theorem.

Theorem 2.7 *$uscore(T, Q)$ is monotonic with respect to $wantf(t, T, Q)$ for any $t \in T$ and $uscore(T, Q) \geq score(T, Q)$ where $score(T, Q)$ is defined in Eq. 2.11.*

Algorithm 9 Skyline-Sweeping (the keyword query Q, the top-k value k, the CN C)

1: $topk \leftarrow \emptyset; Q \leftarrow \emptyset$

2: $Q.push((1, 1, ..., 1), uscore(1, 1, ..., 1))$

3: **while** $Q.max\text{-}uscore > score(topk[k], Q)$ **do**

4:     $c \leftarrow Q.popmax()$

5:     update $topk$ using $eval(c)$

6:     **for** $i = 1$ **to** s **do**

7:         $c' \leftarrow c$

8:         $c'[i] \leftarrow c'[i] + 1$

9:         $Q.push(c', uscore(c'))$

10:        **if** $c'[i] > 1$ **then**

11:            **break**

12: **output** $topk$

Another problem caused by the Global-Pipelined algorithm is that when a new tuple $M_p.cur$ is processed, it tries all the combinations of seen tuples $(t_1, t_2, ..., t_p, t_{p+1}, ..., t_s)$ to test whether each combination can be joined with $M_p.cur$. This operation is costly because the number of combinations can be extremely large when the number of seen tuples becomes large.

The Skyline-Sweeping **Algorithm:** Skyline-Sweeping has been proposed in *SPARK* to handle two problems: (1) dealing with the non-monotonic score function in Eq. 2.11, and (2) significantly reducing the number of combinations tested. Suppose in $M_1, M_2, ..., M_s$ of CN C, tuples are ranked in decreasing order of the *wantf* values. For simplicity, we use $c = (i_1, i_2, ..., i_s)$ to denote the combination of tuples $(M_1.t_{i_1}, M_2.t_{i_2}, ..., M_s.t_{i_s})$ and we use $uscore(i_1, i_2, ..., i_s)$ to denote the *uscore* (Eq. 2.22) for the *MTJNT*s that include tuples $(M_1.t_{i_1}, M_2.t_{i_2}, ..., M_s.t_{i_s})$. The Skyline-Sweeping algorithm is shown in Algorithm 9.

The algorithm processes a single CN C. A priority queue Q is used to keep the set of seen but not tested combinations ordered by *uscore*. Iteratively, a combination c is selected from Q, that has the largest *uscore* (line 4). Every time a combination is selected, it is evaluated to update the *topk* list. Then all of its adjacent combinations are tried in a non-redundant way (lines 6-11), and each adjacent combination is pushed into Q. Lines 10-11 ensure that each combination is enumerated only once. If the maximum score for tuples in Q is no larger than the k-th largest score in the *topk* list, it can stop and output the *topk* list as the final result. The comparison between the processed combinations for the Single-Pipelined algorithm and the processed combinations for the Skyline-Sweeping algorithm is shown in Figure 2.15.

When there are multiple CNs, it can change the Skyline-Sweeping algorithm using the similar methods introduced in the Global-Pipelined algorithm, i.e., it can make Q and $topk$ global to maintain the set of combinations in multiple CNs.

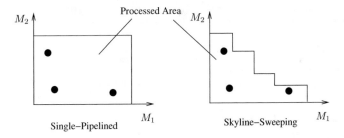

Figure 2.15: Saving computational cost using the Skyline-Sweeping algorithm

The Block-Pipelined **Algorithm:** The upper bound score function in Eq. 2.22 plays two roles in the algorithm: (1) the monotonicity of the upper bound score function ensures that the algorithm can output the correct top-k results when stopping early, (2) It is an estimation of the real score of the results. The tighter the score is, the earlier the algorithm stops. The upper bound score function in Eq. 2.22 may sometimes be very loose, which generates many unnecessary combinations to be tested. In order to decrease such unnecessary combinations, a new Block-Pipelined algorithm is proposed in *SPARK*. A new upper bound score function $bscore$ is introduced, which is tighter than the $uscore$ function in Eq. 2.22, but it is not monotonic. The aim of the Block-Pipelined algorithm is to use both the $uscore$ and the $bscore$ functions such that (1) the $uscore$ function can make sure that the $topk$ results are correctly output, and (2) the $bscore$ function can decrease the gap between the estimated value and the real value of results, and thus reduce the computational cost. The $bscore$ is defined as follows:

$$bscore(T, Q) = bscore_a(T, Q) \cdot score_b(T, Q) \cdot score_c(T, Q) \qquad (2.23)$$

where

$$bscore_a(T, Q) = \sum_{w \in T \cap Q} \frac{1 + \ln(1 + \ln(tf(T, w)))}{1 - s} \cdot \ln(idf(T, w)) \qquad (2.24)$$

The Block-Pipelined algorithm is shown in Algorithm 10; it is similar to the Skyline-Sweeping algorithm. The difference is that it assigns each combination c enumerated a status; for the first time it is enumerated, it calculates its $uscore$, sets its status to be $USCORE$ and inserts it into the queue Q (lines 9-14). Otherwise, if it is already assigned a $USCORE$ status, it calculates its $bscore$, sets its status to be $BSCORE$ and reinserts it into the queue Q again (lines 6-8) before enumerating its neighbors (lines 9-14). If its status is already set to be $BSCORE$, it evaluates it and updates the $topk$ list (line 16). The Block-Pipelined algorithm deals with a single CN case. When there are multiple CNs, it can use the same methods as handling multiple CNs in the Skyline-Sweeping algorithm.

Algorithm 10 Block-Pipelined (the keyword query Q, the top-k value k, the CN C)

1: $topk \leftarrow \emptyset; Q \leftarrow \emptyset$
2: $c \leftarrow (1, 1, ..., 1); c.status = USCORE$
3: $Q.push(c, uscore(c))$
4: **while** $Q.max\text{-}uscore > score(topk[k], Q)$ **do**
5:     $c \leftarrow Q.popmax()$
6:     **if** $c.status = USCORE$ **then**
7:         $c.status = BSCORE$
8:         $Q.push(c, bscore(c))$
9:         **for** $i = 1$ **to** s **do**
10:             $c' \leftarrow c; c'.status = USCORE$
11:             $c'[i] \leftarrow c'[i] + 1$
12:             $Q.push(c', uscore(c'))$
13:             **if** $c'[i] > 1$ **then**
14:                 **break**
15:     **else**
16:         update $topk$ using $eval(c)$
17: **output** $topk$

2.4   OTHER KEYWORD SEARCH SEMANTICS

In the above discussions, for an l-keyword query on a relational database, each result is an *MTJNT*. This is referred to as the *connected tree semantics*. There are two other semantics to answer an l-keyword query on a relational database, namely *distinct root semantics* and *distinct core semantics*. In this section, we will focus on how to answer keyword queries using RDBMS given the schema graph. In the next chapter, we will further discuss how to answer keyword queries under different semantics on a schema free database graph.

Distinct Root Semantics: An l-keyword query finds a collection of tuples that contain all the keywords and that are reachable from a root tuple (center) within a user-given distance (Dmax). The distinct root semantics implies that the same root tuple determines the tuples uniquely [Dalvi et al., 2008; He et al., 2007; Hristidis et al., 2008; Li et al., 2008a; Qin et al., 2009a]. Suppose that there is a result rooted at tuple t_r. For any of the l keywords, say k_i, there is a tuple t in the result that satisfies the following conditions: (1) t contains keyword k_i, (2) among all tuples that contain k_i, the distance between t and t_r is minimum[3], and (3) the minimum distance between t and t_r must be less than or equal to a user given parameter Dmax.

Reconsider the *DBLP* database in Example 2.1 with the same 2-keyword query $Q =$ {Michelle, XML}, and let Dmax = 2. The 10 results are shown in Figure 2.16(a). The root nodes are the nodes shown at the top, and all root nodes are distinct. For example, the rightmost result in

---

[3]If there is a tie, then a tuple is selected with a predefined order among tuples in practice.

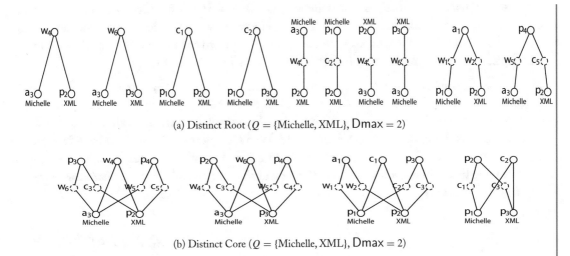

(a) Distinct Root ($Q = \{$Michelle, XML$\}$, $\mathsf{Dmax} = 2$)

(b) Distinct Core ($Q = \{$Michelle, XML$\}$, $\mathsf{Dmax} = 2$)

Figure 2.16: Distinct Root/Core Semantics

Figure 2.16(a) shows that two nodes, a_3 (containing "Michelle") and p_2 (containing "XML"), are reachable from the root node p_4 within $\mathsf{Dmax} = 2$. Under the distinct root semantics, the rightmost result can be output as a set (p_4, a_3, p_2), where the connections from the root node (p_4) to the two nodes can be ignored as discussed in *BLINKS* [He et al., 2007].

Distinct Core Semantics: An l-keyword query finds multi-center subgraphs, called communities [Qin et al., 2009a,b]. A community, $C_i(V, E)$, is specified as follows. V is a union of three subsets of tuples, $V = V_c \cup V_k \cup V_p$, where, V_k is a set of keyword-tuples where a keyword-tuple $v_k \in V_k$ contains at least a keyword, and all l keywords in the given l-keyword query must appear in at least one keyword-tuple in V_k; V_c is a set of center-tuples where there exists at least a sequence of connections between $v_c \in V_c$ and every $v_k \in V_k$ such that $dis(v_c, v_k) \leq \mathsf{Dmax}$; and V_p is a set of path-tuples that appear on a shortest sequence of connections from a center-tuple $v_c \in V_c$ to a keyword-tuple $v_k \in V_k$ if $dis(v_c, v_k) \leq \mathsf{Dmax}$. Note that a tuple may serve several roles as keyword/center/path tuples in a community. E is a set of connections for every pair of tuples in V if they are connected over shortest paths from nodes in V_c to nodes in V_k. A community, C_i, is uniquely determined by the set of keyword tuples, V_k, which is called the core of the community, and denoted as core(C_i).

Reconsider the *DBLP* database in Example 2.1 with the same 2-keyword query $Q = \{$Michelle, XML$\}$ and $\mathsf{Dmax} = 2$. The four communities are shown in Figure 2.16(b), and the four unique cores are (a_3, p_2), (a_3, p_3), (p_1, p_2), and (p_1, p_3), for the four communities from left to right, respectively. The multi-centers for each of the communities are shown in the top. For example, for the rightmost community, the two centers are p_2 and c_2.

It is important to note that the parameter Dmax used in the distinct core/root semantics is different from the parameter Tmax used in the connected tree semantics. Dmax specifies a range from a center (root tuple) in which a tuple containing a keyword can be possibly included in a result, and Tmax specifies the maximum number of nodes to be included in a result.

Distinct Core/Root in RDBMS: We outline the approach to process l-keyword queries with a radius (Dmax) based on the distinct core/root semantics. In the first step, for each keyword k_i, we compute a temporal relation, $Pair_i(tid_i, dis_i, TID)$, with three attributes, where both TID and tid_i are TIDs and dis_i is the shortest distance between TID and tid_i ($dis(TID, tid_i)$), which is less than or equal to Dmax. A tuple in $Pair_i$ indicates that the TID tuple is in the shortest distance of dis_i with the tid_i tuple that contains keyword k_i. In the second step, we join all temporal relations, $Pair_i$, for $1 \leq i \leq l$, on attribute TID (center)

$$ S \leftarrow Pair_1 \underset{Pair_1.TID=Pair_2.TID}{\bowtie} Pair_2 \cdots Pair_{l-1} \underset{Pair_{l-1}.TID=Pair_l.TID}{\bowtie} Pair_l \qquad (2.25) $$

Here, S is a $2l + 1$ attribute relation, $S(TID, tid_1, dis_1, \cdots, tid_l, dis_l)$.

Over the temporal relation S, we can obtain the multi-center communities (distinct core) by grouping tuples on l attributes, $tid_1, tid_2, \cdots, tid_l$. Consider query $Q = \{Michelle, XML\}$ and Dmax $= 2$, against the simple *DBLP* database in Figure 2.2. The rightmost community in Figure 2.16(b) is shown in Figure 2.17.

TID	tid1	dis1	tid2	dis2
p_1	p_1	0	p_3	2
p_2	p_1	2	p_3	2
p_3	p_1	2	p_3	0
c_2	p_1	1	p_3	1

Figure 2.17: A Multi-Center Community

Here, the distinct core consists of p_1 and p_3, where p_1 contains keyword "Michelle" (k_1) and p_3 contains keyword "XML" (k_2), and the four centers, $\{p_1, p_2, p_3, c_2\}$, are listed in the TID column. Any center can reach all the tuples in the core, $\{p_1, p_3\}$, within Dmax. The above does not explicitly include the two nodes, c_1 and c_3 in the rightmost community in Figure 2.16(b), which can be maintained in an additional attribute by concatenating the TIDs, for example, $p_2.c_1.p_1$ and $p_2.c_3.p_3$. In a similar fashion, over the same temporal relation S, we can also obtain the distinct root results by grouping tuples on the attribute TID. Consider the query $Q = \{Michelle, XML\}$ and Dmax $= 2$, the rightmost result in Figure 2.16(a) is shown in Figure 2.18.

The distinct root is represented by the TID, and the rightmost result in Figure 2.16(a) is the first of the two tuples, where a_3 contains keyword "Michelle" (k_1) and p_2 contains keyword "XML"

TID	tid1	dis1	tid2	dis2
p_4	a_3	2	p_2	2
p_4	a_3	2	p_3	2

Figure 2.18: A Distinct Root Result

Gid	TID	tid1	dis1	tid2	dis2
1	a_3	a_3	0	p_2	2
1	w_4	a_3	1	p_2	1
1	p_2	a_3	2	p_2	0
1	p_3	a_3	2	p_2	2
1	p_4	a_3	2	p_2	2
2	a_3	a_3	0	p_3	2
2	w_6	a_3	1	p_3	1
2	p_2	a_3	2	p_3	2
2	p_3	a_3	2	p_3	0
2	p_4	a_3	2	p_3	2
3	a_1	p_1	2	p_2	2
3	p_1	p_1	0	p_2	2
3	p_2	p_1	2	p_2	0
3	p_3	p_1	2	p_2	2
3	c_1	p_1	1	p_2	1
4	p_1	p_1	0	p_3	2
4	p_2	p_1	2	p_3	2
4	p_3	p_1	2	p_3	0
4	c_2	p_1	1	p_3	1

Gid	TID	tid1	dis1	tid2	dis2
1	w_4	a_3	1	p_2	1
2	w_6	a_3	1	p_3	1
3	c_1	p_1	1	p_2	1
4	c_2	p_1	1	p_3	1
5	a_3	a_3	0	p_2	2
5	a_3	a_3	0	p_3	2
6	p_1	p_1	0	p_2	2
6	p_1	p_1	0	p_3	2
7	p_2	a_3	2	p_2	0
7	p_2	a_3	2	p_3	2
7	p_2	p_1	2	p_2	0
7	p_2	p_1	2	p_3	2
8	p_3	a_3	2	p_3	0
8	p_3	a_3	2	p_2	2
8	p_3	p_1	2	p_2	2
8	p_3	p_1	2	p_3	0
9	a_1	p_1	2	p_2	2
10	p_4	a_3	2	p_2	2
10	p_4	a_3	2	p_3	2

Figure 2.19: Distinct Core(left) and Distinct Root(right) (Q = {Michelle, XML}, Dmax = 2)

(k_2). Note that a distinct root means a result is uniquely determined by the root. As shown above, there are two tuples with the same root p_4. We select one of them using the aggregate function min.

The complete results for the distinct core/root results are given in Figure 2.19, for the same 2-keyword query, Q = {Michelle, XML} with Dmax = 2, against the *DBLP* database in Figure 2.2. Both tables have an attribute Gid that is for easy reference of the distinct core/root results. The left table shows the same content as the right table by grouping on TID in which the shadowed tuples are removed using the SQL aggregate function min to ensure the distinct root semantics.

Naive Algorithms: Figure 2.20 outlines the two main steps for processing the distinct core/root 2-keyword query, Q = {Michelle, XML}, with Dmax = 2 against the simple *DBLP* database. Its schema graph, G_S, is in Figure 2.1, and the database is in Figure 2.2. In Figure 2.20, the left side computes $Pair_1$ and $Pair_2$ temporal relations, for keyword k_1 = "Michelle" and k_2 = "XML", using

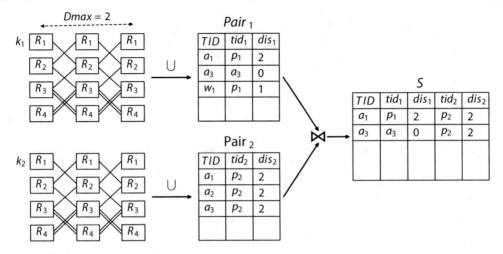

Figure 2.20: An Overview (R_1, R_2, R_3, and R_4 represent Author, Write, Paper, and Cite relations in Example 2.1)

projects, joins, unions, and group-by, and the right side joins $Pair_1$ and $Pair_2$ to compute the S relation (Eq. 2.25).

Let R_1, R_2, R_3, and R_4 represent Author, Write, Paper, and Cite relations. The $Pair_1$ for the keyword k_1 is produced in the following steps.

$$
\begin{aligned}
P_{0,1} &\leftarrow \Pi_{TID \rightarrow tid_1, 0 \rightarrow dis_1, *}(\sigma_{contain(k_1)} R_1) \\
P_{0,2} &\leftarrow \Pi_{TID \rightarrow tid_1, 0 \rightarrow dis_1, *}(\sigma_{contain(k_1)} R_2) \\
P_{0,3} &\leftarrow \Pi_{TID \rightarrow tid_1, 0 \rightarrow dis_1, *}(\sigma_{contain(k_1)} R_3) \\
P_{0,4} &\leftarrow \Pi_{TID \rightarrow tid_1, 0 \rightarrow dis_1, *}(\sigma_{contain(k_1)} R_4)
\end{aligned} \tag{2.26}
$$

Here $\sigma_{contain(k_1)} R_j$ selects the tuples in R_j that contain the keyword k_1. Let $R'_j \leftarrow \sigma_{contain(k_1)} R_j$, $\Pi_{TID \rightarrow tid_1, 0 \rightarrow dis_1, *}(R'_j)$ projects tuples from R'_j with all attributes ($*$) by further adding two attributes (renaming the attribute TID to be tid_1 and adding a new attribute dis_1 with an initial value zero (this is supported in SQL)). For example, $\Pi_{TID \rightarrow tid_1, 0 \rightarrow dis_1, *}(\sigma_{contain(k_1)} R_1)$ is translated into the following SQL.

select TID **as** tid_1, 0 **as** dis_1, TID, Name **from** Author **as** R_1 **where** contain(Title, Michelle)

The meaning of the temporal relation $P_{0,1}(tid_1, dis_1, TID, Name)$ is a set of R_1 relation tuples (identified by TID) that are in distance $dis_1 = 0$ from the tuples (identified by tid_1) containing keyword k_1 = "Michelle". The same is true for other $P_{0,j}$ temporal relations as well. After $P_{0,j}$ are computed, $1 \le j \le 4$, we compute $P_{1,j}$ followed by $P_{2,j}$ to obtain R_j relation tuples that are in distance 1 and distance 2 from the tuples containing keyword k_1 = "Michelle" (Dmax = 2). Note that relation $P_{d,j}$ contains the set of tuples of R_j that are in distance d from a tuple containing a certain keyword, and its two attributes, tid_l and dis_l, explicitly indicate that it is about keyword k_l. The details of computing $P_{1,j}$ for R_j, $1 \leq j \leq 4$, are given below.

$$P_{1,1} \leftarrow \Pi_{P_{0,2}.TID \rightarrow tid_1, 1 \rightarrow dis_1, R_1.*}(P_{0,2} \underset{P_{0,2}.AID = R_1.TID}{\bowtie} R_1)$$

$$P_{1,2} \leftarrow \Pi_{P_{0,1}.TID \rightarrow tid_1, 1 \rightarrow dis_1, R_2.*}(P_{0,1} \underset{P_{0,1}.TID = R_2.AID}{\bowtie} R_2) \cup$$

$$\Pi_{P_{0,3}.TID \rightarrow tid_1, 1 \rightarrow dis_1, R_2.*}(P_{0,3} \underset{P_{0,3}.TID = R_2.PID}{\bowtie} R_2)$$

$$P_{1,3} \leftarrow \Pi_{P_{0,2}.TID \rightarrow tid_1, 1 \rightarrow dis_1, R_3.*}(P_{0,2} \underset{P_{0,2}.PID = R_3.TID}{\bowtie} R_3) \cup$$

$$\Pi_{P_{0,4}.TID \rightarrow tid_1, 1 \rightarrow dis_1, R_3.*}(P_{0,4} \underset{P_{0,4}.PID1 = R_3.TID}{\bowtie} R_3) \cup$$

$$\Pi_{P_{0,4}.TID \rightarrow tid_1, 1 \rightarrow dis_1, R_3.*}(P_{0,4} \underset{P_{0,4}.PID2 = R_3.TID}{\bowtie} R_3)$$

$$P_{1,4} \leftarrow \Pi_{P_{0,3}.TID \rightarrow tid_1, 1 \rightarrow dis_1, R_4.*}(P_{0,3} \underset{P_{0,3}.TID = R_4.PID1}{\bowtie} R_4) \cup$$

$$\Pi_{P_{0,3}.TID \rightarrow tid_1, 1 \rightarrow dis_1, R_4.*}(P_{0,3} \underset{P_{0,3}.TID = R_4.PID2}{\bowtie} R_4) \tag{2.27}$$

Here, each join/project corresponds to a foreign key reference – an edge in schema graph G_S. The idea is to compute $P_{d,j}$ based on $P_{d-1,i}$ if there is an edge between R_j and R_i in G_S. Consider $P_{1,3}$ for R_3, it computes $P_{1,3}$ by union of three joins ($P_{0,2} \bowtie R_3 \cup P_{0,4} \bowtie R_3 \cup P_{0,4} \bowtie R_3$), because there is one foreign key reference between R_3 (Paper) and R_2 (Write), and two foreign key references between R_3 and R_4 (Cite). This ensures that all R_j tuples that are with distance d from a tuple containing a keyword k_l can be computed. Continuing the example, to compute $P_{2,j}$ for R_j, $1 \leq j \leq 4$, for keyword k_1, we replace every $P_{d,j}$ in Eq. 2.27 with $P_{d+1,j}$ and replace "$1 \rightarrow dis_1$" with "$2 \rightarrow dis_1$". The process repeats Dmax times.

Suppose that we have computed $P_{d,j}$ for $0 \leq d \leq$ Dmax and $1 \leq j \leq 4$, for keyword $k_1 =$ "Michelle". We further compute the shortest distance between a R_j tuple and a tuple containing k_1 using union, group-by, and SQL aggregate function min. First, we perform project, $P_{d,j} \leftarrow \Pi_{TID, tid_1, dis_1} P_{d,j}$. Therefore, every $P_{d,j}$ relation has the same tree attributes. Second, for R_j, we compute the shortest distance from a R_j tuple to a tuple containing keyword k_1 using group-by (Γ) and SQL aggregate function min.

$$G_j \leftarrow_{TID, tid_1} \Gamma_{\min(dis_1)}(P_{0,j} \cup P_{1,j} \cup P_{2,j}) \tag{2.28}$$

where, the left side of group-by (Γ) is group-by attributes, and the right side is the SQL aggregate function. Finally,

$$Pair_1 \leftarrow G_1 \cup G_2 \cup G_3 \cup G_4 \tag{2.29}$$

Here, $Pair_1$ records all tuples that are shortest distance away from a tuple containing keyword k_1, within Dmax. Note that $G_i \cap G_j = \emptyset$, because G_i and G_j are tuples identified with TIDs from R_i and R_j relations and TIDs are unique in the database as assumed. We can compute $Pair_2$ for keyword $k_2 =$ "XML" following the same procedure as indicated in Eq. 2.26-Eq. 2.29. Once all $Pair_1$ and $Pair_2$ are computed, we can easily compute distinct core/root results based on the relation $S \leftarrow Pair_1 \bowtie Pair_2$ (Eq. 2.25).

Algorithm 11 $Pair(G_S, k_i, \text{Dmax}, R_1, \cdots, R_n)$

Input: Schema G_S, keyword k_i, Dmax, n relations $R_1, \cdots, R_n$.
Output: $Pair_i$ with 3 attributes: TID, tid_i, dis_i.

1: **for** $j = 1$ **to** n **do**
2:    $P_{0,j} \leftarrow \prod_{R_j.TID \to tid_i,\ 0 \to dis_i,\ R_j.*}(\sigma_{contain(k_i)} R_j)$
3:    $G_j \leftarrow \prod_{tid_i, dis_i, TID}(P_{0,j})$
4: **for** $d = 1$ **to** Dmax **do**
5:    **for** $j = 1$ **to** n **do**
6:       $P_{d,j} \leftarrow \emptyset$
7:       **for all** $(R_j, R_l) \in E(G_S) \vee (R_l, R_j) \in E(G_S)$ **do**
8:          $\Delta \leftarrow \prod_{P_{d-1,l}.TID \to tid_i,\ d \to dis_i,\ R_j.*}(P_{d-1,l} \bowtie R_j)$
9:          $\Delta \leftarrow \sigma_{(tid_i, TID) \notin \prod_{tid_i, TID}(G_j)}(\Delta)$
10:         $P_{d,j} \leftarrow P_{d,j} \cup \Delta$
11:         $G_j \leftarrow G_j \cup \prod_{tid_i, dis_i, TID}(\Delta)$
12: $Pair_i \leftarrow G_1 \cup G_2 \cup \cdots \cup G_n$
13: **return** $Pair_i$

Computing group-by (Γ) with SQL aggregate function min: Consider Eq. 2.28, the group-by Γ can be computed by virtually pushing Γ. Recall that all $P_{d,j}$ relations, for $1 \leq d \leq \text{Dmax}$, have the same schema, and $P_{d,j}$ maintains R_j tuples that are in distance d from a tuple containing a keyword. We use two pruning rules to reduce the number of temporal tuples computed.

Rule-1: If the same (tid_i, TID) value appears in two different $P_{d',j}$ and $P_{d,j}$, then the shortest distance between tid_i and TID must be in $P_{d',j}$ but not $P_{d,j}$, if $d' < d$. Therefore, Eq. 2.28 can be computed as follows.

$$
\begin{aligned}
G_j &\leftarrow P_{0,j} \\
G_j &\leftarrow G_j \cup (\sigma_{(tid_1, TID) \notin \prod_{tid_1, TID}(G_j)} P_{1,j}) \\
G_j &\leftarrow G_j \cup (\sigma_{(tid_1, TID) \notin \prod_{tid_1, TID}(G_j)} P_{2,j})
\end{aligned}
\tag{2.30}
$$

Here, $\sigma_{(tid_1, TID) \notin \prod_{tid_1, TID}(G_j)} P_{2,j}$ selects $P_{2,j}$ tuples where their (tid_1, TID) does not appear in G_j; yet, in other words, there does not exist a shortest path between tid_1 and TID before.

Rule-2: If there exists a shortest path between tid_i and TID value pair, say, $dis_i(tid_i, TID) = d'$, then there is no need to compute any tuple connections between the tid_i and TID pair, because all those will be removed later by group-by and SQL aggregate function min. In Eq. 2.27, every $P_{1,j}$, $1 \leq j \leq 4$, can be further reduced as $P_{1,j} \leftarrow \sigma_{(tid_1, TID) \notin \prod_{tid_1, TID}(P_{0,j})} P_{1,j}$.

    The algorithm $Pair()$ is given in Algorithm 11, which computes $Pair_i$ for keyword k_i. It first computes all the initial $P_{0,j}$ relations (refer to Eq. 2.26) and initializes G_j relations (refer to the first equation in Eq. 2.30) in lines 1-3. Second, it computes $P_{d,j}$ for every $1 \leq d \leq \text{Dmax}$ and every

---

Algorithm 12 $DC\text{-}Naive(R_1, \cdots, R_n, G_S, Q, \text{Dmax})$

Input: n relations $R_1, R_2, \cdots, R_n$, schema graph G_S, and
         l-keyword, $Q = \{k_1, k_2, \cdots, k_l\}$, and radius Dmax.
Output:Relation with $2l + 1$ attributes named $TID, tid_1, dis_1, \cdots, tid_l, dis_l$.

1: **for** $i = 1$ **to** l **do**
2:     $Pair_i \leftarrow Pair(G_S, k_i, \text{Dmax}, R_1, \cdots, R_n)$
3: $S \leftarrow Pair_1 \bowtie Pair_2 \bowtie \cdots \bowtie Pair_l$
4: Sort S by $tid_1, tid_2, \cdots, tid_l$
5: **return** S

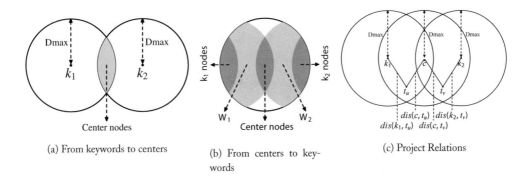

(a) From keywords to centers   (b) From centers to keywords   (c) Project Relations

Figure 2.21: Three-Phase Reduction

relation R_j, $1 \le j \le n$, in two "for loops" (lines 4-5). In lines 7-11, it computes $P_{d,j}$ based on the foreign key references in the schema graph G_S, referencing to Eq. 2.27 and Eq. 2.30, using the two rules, Rule-1 and Rule-2. In our example, to compute $Pair_1$, it calls $Pair(G_S, k_1, \text{Dmax}, R_1, R_2, R_3, R_4)$, where k_1 = "Michelle", Dmax $= 2$, and the 4 relations R_j, $1 \le j \le 4$.

The naive algorithm $DC\text{-}Naive()$ to compute distinct cores is outlined in Algorithm 12. $DR\text{-}Naive()$ that computes distinct roots can be implemented in the same way as $DC\text{-}Naive()$ by replacing line 4 in Algorithm 12 with 2 group-bys as follows: $X \leftarrow {}_{TID}\Gamma_{\min(dis_1) \rightarrow dis_1, \cdots, \min(dis_l) \rightarrow dis_l} S$, and $S \leftarrow {}_{TID, dis_1, \cdots, dis_l}\Gamma_{\min(tid_1) \rightarrow tid_1, \cdots, \min(tid_l) \rightarrow tid_l}(S \bowtie X)$.

Three-Phase Database Reduction: We now discuss a three-phase reduction approach to project a relational database RDB' out of RDB with which we compute multi-center communities (distinct core semantics). In other words, in the three-phase reduction, we significantly prune the tuples from an RDB that do not participate in any communities. We also show that we can fast compute distinct root results using the same subroutine used in the three-phase reduction.

Algorithm 13 $DC(R_1, R_2, \cdots, R_n, G_S, Q, \text{Dmax})$

Input: n relations $R_1, R_2, \cdots, R_n$, with schema graph G_S, and
an l-keyword query, $Q = \{k_1, k_2, \cdots, k_l\}$, and radius Dmax.

Output: Relation with $2l + 1$ attributes named TID, $tid_1, dis_1, \cdots, tid_l, dis_l$.

1: **for** $i = 1$ **to** l **do**
2:     $\{G_{1,i}, \cdots, G_{n,i}\} \leftarrow PairRoot(G_S, k_i, \text{Dmax}, R_1, \cdots, R_n, \sigma_{contain(k_i)} R_1, \cdots, \sigma_{contain(k_i)} R_n)$
3:     **for** $j = 1$ **to** n **do**
4:        $R_{j,i} \leftarrow R_j \ltimes G_{j,i}$
5: **for** $j = 1$ **to** n **do**
6:     $Y_j \leftarrow G_{j,1} \bowtie G_{j,2} \bowtie \cdots \bowtie G_{j,l}$
7:     $X_j \leftarrow R_j \ltimes Y_j$
8: **for** $i = 1$ **to** l **do**
9:     $\{W_{1,i}, \cdots, W_{n,i}\} \leftarrow PairRoot(G_S, k_i, \text{Dmax}, R_{1,i}, \cdots, R_{n,i}, X_1, \cdots, X_n)$
10:     **for** $j = 1$ **to** n **do**
11:        $Path_{j,i} \leftarrow G_{j,i} \bowtie_{G_{j,i}.TID = W_{j,i}.TID} W_{j,i}$
12:        $Path_{j,i} \leftarrow \prod_{TID, G_{j,i}.dis_i \to d_{k_i}, W_{j,i}.dis_i \to d_r} (Path_{j,i})$
13:        $Path_{j,i} \leftarrow \sigma_{d_{k_i} + d_r \leq \text{Dmax}} (Path_{j,i})$
14:        $R'_{j,i} \leftarrow R_{j,i} \ltimes Path_{j,i}$
15: **for** $i = 1$ **to** l **do**
16:     $Pair_i \leftarrow Pair(R'_{1,i}, R'_{2,i}, \cdots, R'_{n,i}, G_S, k_i, \text{Dmax})$
17: $S \leftarrow Pair_1 \bowtie Pair_2 \bowtie \cdots \bowtie Pair_l$
18: Sort S by $tid_1, tid_2, \cdots, tid_l$
19: **return** S

Figure 2.21 outlines the main ideas for processing an l-keyword query, $Q = \{k_1, k_2, \cdots, k_l\}$, with a user-given Dmax, against an *RDB* with a schema graph G_S.

The first reduction phase (from keyword to center): We consider a keyword k_i as a virtual node, called a keyword-node, and we take a keyword-node, k_i, as a center to compute all tuples in an *RDB* that are reachable from k_i within Dmax. A tuple t within Dmax from a virtual keyword-node k_i means that tuple t can reach at least a tuple containing k_i within Dmax. Let $\mathcal{G}_i$ be the set of tuples in *RDB* that can reach at least a tuple containing keyword k_i within Dmax, for $1 \leq i \leq l$. Based on all $\mathcal{G}_i$, we can compute $\mathcal{Y} = \mathcal{G}_1 \bowtie \mathcal{G}_2 \bowtie \cdots \bowtie \mathcal{G}_l$, which is the set of center-nodes that can reach every keyword-node k_i, $1 \leq i \leq l$, within Dmax. $\mathcal{Y}$ is illustrated as the shaded area in Figure 2.21(a) for $l = 2$. Obviously, a center appears in a multi-center community must appear in $\mathcal{Y}$.

The second reduction phase (from center to keyword): In a similar fashion, we consider a virtual center-node. A tuple t within Dmax from a virtual center-node means that t is reachable from a tuple in $\mathcal{Y}$ within Dmax. We compute all tuples that are reachable from $\mathcal{Y}$ within Dmax. Let $\mathcal{W}_i$

Algorithm 14 $PairRoot(G_S, k_i, \text{Dmax}, R_1, \cdots, R_n, I_1, \cdots, I_n)$

Input: Schema graph G_S, keyword k_i, Dmax, n relations $R_1, R_2, \cdots, R_n$, and n initial
relations $I_1, I_2, \cdots, I_n$.

Output: n relations $G_{1,i}, \cdots, G_{n,i}$ each has 3 attributes: TID, tid_i, dis_i.

1: **for** $j = 1$ **to** n **do**
2: $\quad P_{0,j} \leftarrow \prod_{I_j.TID \rightarrow tid_i, 0 \rightarrow dis_i, I_j.*}(I_j)$
3: $\quad G_{j,i} \leftarrow \prod_{tid_i, dis_i, TID}(P_{0,j})$
4: **for** $d = 1$ **to** Dmax **do**
5: $\quad$ **for** $j = 1$ **to** n **do**
6: $\quad\quad P_{d,j} \leftarrow \emptyset$
7: $\quad\quad$ **for all** $(R_j, R_l) \in E(G_S) \vee (R_l, R_j) \in E(G_S)$ **do**
8: $\quad\quad\quad \Delta \leftarrow \prod_{P_{d-1,l}.TID \rightarrow tid_i, d \rightarrow dis_i, R_j.*}(P_{d-1,l} \bowtie R_j)$
9: $\quad\quad\quad \Delta \leftarrow_{R_j.*} \Gamma_{\min(tid_i), \min(dis_i)}(\Delta)$
10: $\quad\quad\quad \Delta \leftarrow \sigma_{TID \notin \prod_{TID}(G_{j,i})}(\Delta)$
11: $\quad\quad\quad P_{d,j} \leftarrow P_{d,j} \cup \Delta$
12: $\quad\quad\quad G_{j,i} \leftarrow G_{j,i} \cup \prod_{tid_i, dis_i, TID}(\Delta)$
13: **return** $\{G_{1,i}, \cdots, G_{n,i}\}$

Algorithm 15 $DR(R_1, R_2, \cdots, R_n, G_S, Q, \text{Dmax})$

Input: n relations $R_1, R_2, \cdots, R_n$, with schema graph G_S, and
an l-keyword query, $Q = \{k_1, k_2, \cdots, k_l\}$, and radius Dmax.

Output: Relation with $2l + 1$ attributes named TID, $tid_1, dis_1, \cdots, tid_l, dis_l$.

1: **for** $i = 1$ **to** l **do**
2: $\quad \{G_{1,i}, \cdots, G_{n,i}\} \leftarrow PairRoot(G_S, k_i, \text{Dmax}, R_1, \cdots, R_n, \sigma_{contain(k_i)}R_1, \cdots, \sigma_{contain(k_i)}R_n)$
3: **for** $j = 1$ **to** n **do**
4: $\quad S_j \leftarrow G_{j,1} \bowtie G_{j,2} \bowtie \cdots \bowtie G_{j,l}$
5: $S \leftarrow S_1 \cup S_2 \cup \cdots \cup S_n$
6: **return** S

---

be the set of tuples in $\mathcal{G}_i$ that can be reached from a center in $\mathcal{Y}$ within Dmax, for $1 \leq i \leq l$. Note that $\mathcal{W}_i \subseteq \mathcal{G}_i$. When $l = 2$, $\mathcal{W}_1$ and $\mathcal{W}_2$ are illustrated as the shaded areas on left and right in Figure 2.21(b), respectively. Obviously, only the tuples that contain a keyword within Dmax from a center are possible to appear in the final result as keyword tuples.

The third reduction phase (project DB): We project an RDB' out of the RDB, which is sufficient to compute all multi-center communities by join $\mathcal{G}_i \bowtie \mathcal{W}_i$, for $1 \leq i \leq l$. Consider a tuple in $\mathcal{G}_i$, which contains a TID t' with a distance to the virtual keyword-node k_i, denoted as $dis(t', k_i)$, and consider a tuple in $\mathcal{W}_i$, which contains a TID t' with a distance to the virtual center-node c, denoted as $dis(t', c)$. If $dis(t', k_i) + dis(t', c) \leq$ Dmax, the tuple t' will be projected from the *RDB*. Here, both $dis(t', c)$ and $dis(t', k_i)$ are in the range of [0, Dmax]. In this phase, all such tuples, t', will be projected, which are sufficient to compute all multi-center communities, because the set of such tuples contain every keyword-tuple, center-tuple, and path-tuple to compute all communities. This is illustrated in Figure 2.21(c), when $l = 2$.

The new *DC*() algorithm to compute communities under distinct core semantics is given in Algorithm 13. Suppose that there are n relations in an *RDB* for an l-keyword query. The first reduction phase is in lines 1-7. The second/third reduction phases are done in a for-loop (lines 8-14) in which the second reduction phase is line 9, and the third reduction phase is in lines 10-14. Lines 15-17 are similar as done in *DC-Naive*() to compute communities using $Pair_i$, $1 \leq i \leq l$, and S relation. For the first reduction, it computes $G_{j,i}$ for every keyword k_i and every relation R_j separately by calling a procedure *PairRoot*() (Algorithm 14). *PairRoot*() is designed in a similar fashion to *Pair*(). The main difference is that *PairRoot*() computes tuples, t, that are in shortest distance to a virtual node (keyword or center) within Dmax. Take keyword-nodes as an example. The shortest distance to a tuple containing a keyword is more important than which tuple contains a keyword. Therefore, we only maintain the shortest distance (line 9 in Algorithm 14). *PairRoot*() returns a collection of $G_{j,i}$, for a given keyword k_i, for $1 \leq j \leq n$. Note that $\mathcal{G}_i = \bigcup_{j=1}^{n} G_{j,i}$. In lines 3-4, it projects R_j using semijoin $R_{j,i} \leftarrow R_j \ltimes G_{j,i}$. Here, $R_{j,i} (\subseteq R_j)$ is a set of tuples that are within Dmax from a virtual keyword-node k_i. Note that $\mathcal{Y} = \bigcup_{j=1}^{n} Y_j$. $X_j (\subseteq R_j)$ is a set of centers in relation R_j (line 7). In line 9, starting from all center nodes $(X_1, \cdots, X_n)$, it computes $W_{j,i}$, for keyword k_i, for $1 \leq j \leq n$. Note that $\mathcal{W}_i = \bigcup_{j=1}^{n} W_{j,i}$. In lines 10-14, it further projects $R'_{j,i}$ out of $R_{j,i}$, for a keyword k_i, for $1 \leq j \leq n$. In line 16, it computes $Pair_i$, using the projected relations, $R'_{1,i}, R'_{2,i}, \cdots, R'_{n,i}$. The new algorithm *DR*() to compute distinct roots is given in Algorithm 15.

CHAPTER 3

Graph-Based Keyword Search

In this chapter, we show how to answer keyword queries on a general data graph using graph algorithms. It is worth noting that an *XML* database or World Wide Web also can be modeled as a graph, and there also exists general graphs with textual information stored on the nodes. In the previous chapter, we discussed keyword search on a relational database (*RDB*) using the underlying relational schema that specifies how tuples are connected to each other. Based on the primary and foreign key references defined on a relational schema, an *RDB* can be modeled as a data graph where nodes represent tuples and edges represent the foreign key references between tuples.

In Section 3.1, we discuss graph models and define the problem, precisely. In Section 3.2, we introduce two algorithms that will be used in the subsequent discussions. One is polynomial delay and the other is Dijkstra's single source shortest path algorithm. In Section 3.3, we discuss several algorithms that find Steiner trees as answers for *l*-keyword queries. We will discuss exact and approximate algorithms in Section 3.3. In Section 3.4, we discuss algorithms that find tree-structured answers which have a distinct root. Some indexing approaches and algorithms that deal with external graphs on disk will be discussed. In Section 3.5, we discuss algorithms that find subgraphs.

3.1 GRAPH MODEL AND PROBLEM DEFINITION

Abstract directed weighted graph: As an abstraction, we consider a general directed graph in this chapter, $G_D(V, E)$, where edges have weight $w_e(\langle u, v \rangle)$. For an undirected graph, backward edges with the same weights can be added to make it to be a directed graph. In some definitions, the nodes also have weights to reflect the prestige like the PageRank value [Brin and Page, 1998]. But the algorithms remain the same with little modifications, so we will assume that only edges have weights for the ease of presentation. We use $V(G)$ and $E(G)$ to denote the set of nodes and the set of edges for a given graph G, respectively. We also denote the number of nodes and the number of edges in graph G, using $n = |V(G)|$ and $m = |E(G)|$. In the following, we discuss how to model an *RDB* and *XML* database as a graph, and how weights are assigned to edges.

The (structure and textual) information stored in an *RDB* can be captured by a weighted directed graph, $G_D = (V, E)$. Each tuple t_v in *RDB* is modeled as a node $v \in V$ in G_D, associated with keywords contained in the corresponding tuple. For any two nodes $u, v \in V$, there is a directed edge $\langle u, v \rangle$ (or $u \to v$) if and only if there exists a foreign key on tuple t_u that refers to the primary key in tuple t_v. This can be easily extended to other types of connections; for example, the model can be extended to include edges corresponding to inclusion dependencies [Bhalotia et al., 2002],

TID	Code	Name	Capital	Government
t_1	B	Belgium	BRU	Monarchy
t_2	NOR	Norway	OSL	Monarchy

(a) Countries

TID	Name	Headq	#members
t_3	EU	BRU	25
t_4	ESA	PAR	17

(b) Organizations

TID	Code	Name	Country	Population
t_5	ANT	Antwerp	B	455,148
t_6	BRU	Brussels	B	141,312
t_7	OSL	Oslo	NOR	533,050

(c) Cities

TID	Country	Organization
t_8	B	ESA
t_9	B	EU
t_{10}	NOR	ESA

(d) Members

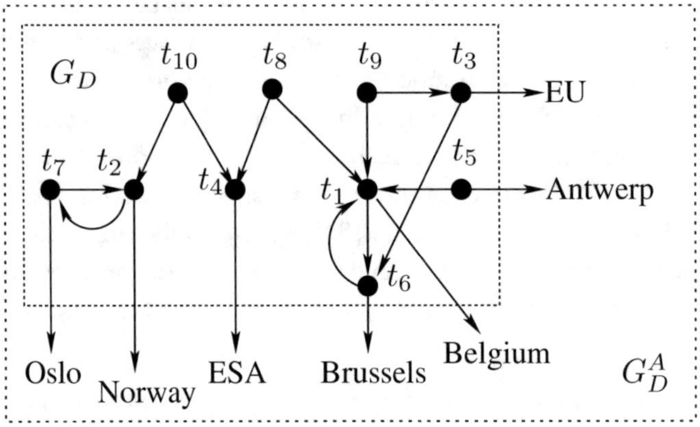

(e) Data Graph

Figure 3.1: A small portion of the Mondial *RDB* and its data graph [Golenberg et al., 2008]

where the values in the referencing column of the referencing relation are contained in the referred column of the referred relation, but the referred column need not to be a key of the referred relation.

Example 3.1   Figure 3.1 shows a small portion of the Mondial relational database. The Name attributes of the first three relations store text information where keywords can be matched. The directed graph transformed from the *RDB*, G_D is depicted in the dotted rectangle in Figure 3.1(e). In Figure 3.1(e), there are keyword nodes for all words appearing in the text attribute of the database. The edge from t_i to keyword node w_j means that the node t_i contains word w_j.

Weights are assigned to edges to reflect the (directional) proximity of the corresponding tuples, denoted as $w_e(\langle u, v \rangle)$. A commonly used weighting scheme [Bhalotia et al., 2002; Ding et al., 2007] is as follows. For a foreign key reference from t_u to t_v, the weight for the directed edge $\langle u, v \rangle$ is given as Eq. 3.1, and the weight for the backward edge $\langle v, u \rangle$ is given as Eq. 3.2.

$$w_e(\langle u, v \rangle) = 1 \qquad (3.1)$$
$$w_e(\langle v, u \rangle) = \log_2(1 + N_{in}(v)) \qquad (3.2)$$

where $N_{in}(v)$ is the number of tuples that refer to t_v, which is the tuple corresponding to node v.

An *XML* document can be naturally represented as a directed graph. Each element is modeled as a node, the sub-element relationships and ID/IDREF reference relationships are modeled as directed edges. One possible weighting scheme [Golenberg et al., 2008] is as follows. First, consider the edges corresponding to sub-element relationship. Let $out(v \rightarrow t)$ denote the number of edges that lead from v to nodes that have the tag t. Similarly, $in(t \rightarrow v)$ denotes the number of edges that lead to v from nodes with tag t. The weight of an edge $\langle v_1, v_2 \rangle$, where the tags of v_1 and v_2 are t_1 and t_2, respectively, is defined as follows.

$$w_e(\langle v_1, v_2 \rangle) = \log(1 + \alpha \cdot out(v_1 \rightarrow t_2) + (1 - \alpha) \cdot in(t_1 \rightarrow v_2))$$

The general idea is that the edges carry more information if there are a few edges that emanate from v_1 and lead to nodes that have the same tag as v_2, or a few edges that enter v_2 and emanate from nodes with the same tag as v_1. The weight of edges that correspond to ID references are set to 0, as they represent strong semantic connections.

The web can also be modelled as a directed graph [Li et al., 2001], $G_D = (V, E)$, where V is the set of physical pages, and E is the hyper- or semantic-links connecting these pages. For a keyword query, it finds connected trees called "information unit," which can be viewed as a logical web document consisting of multiple physical pages as one atomic retrieval unit. Other databases, e.g., RDF and OWL, which are two major W3C standards in semantic web, also conform to the node-labeled graph models.

Given a directed weighted data graph G_D, an l-keyword query consists of a set of $l \geq 2$ keywords, i.e., $Q = \{k_1, k_2, \cdots, k_l\}$. The problem is to find a set of subgraphs of G_D, $\mathcal{R}(G_D, Q) = \{R_1(V, E), R_2(V, E), \cdots\}$, where each $R_i(V, E)$ is a connected subgraph of G_D that contains all the l keywords. Different requirements for the property of subgraphs that should be returned have been proposed in the literature. There are mainly two different structural requirements: (1) a reduced tree that contains all the keywords that we refer to as *tree-based semantics*; (2) a subgraph, such as r-radius steiner graph [Li et al., 2008a], and multi-center induced graph [Qin et al., 2009b]; we call this *subgraph-based semantics*. In the following, we show the tree-based semantics, and we will study the subgraph-based semantics in Section 3.5 in detail.

Tree Answer: In the tree-based semantics, an answer to Q (called a Q-SUBTREE) is defined as any subtree T of G_D that is reduced with respect to Q. Formally, there exists a sequence of l nodes in T, $\langle v_1, \cdots, v_l \rangle$ where $v_i \in V(T)$ and v_i contains keyword term k_i for $1 \leq i \leq l$, such that the leaves of T can only come from those nodes, i.e., $leaves(T) \subseteq \{v_1, v_2, \cdots, v_l\}$, the root of T should also be from those nodes if it has only one child, i.e., $root(T) \in \{v_1, v_2, \cdots, v_l\}$.

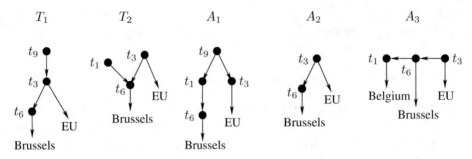

Figure 3.2: Subtrees [Golenberg et al., 2008]

Example 3.2   Consider the five subgraphs in Figure 3.2. Let's ignore all the leave nodes (which are keyword nodes), four of them are directed rooted subtrees, namely T_1, A_1, A_2 and A_3, and the subgraph T_2 is not a directed rooted subtree. For a 2-keyword query $Q = \{$Brussels, EU$\}$, (1) T_1 is not a Q-SUBTREE, because the root t_9 has only one child and t_9 does not contain any keywords, (2) A_1 is a Q-SUBTREE, (3) A_2 is also a Q-SUBTREE, although the root t_3 has only one child, t_3 contains a keyword "EU". Subtree A_3 is not a Q-SUBTREE for Q, but it is for query $Q' = \{$Belgium, Brussels, EU$\}$.

From the above definition of a tree answer, it is not intuitive to distinguish a Q-SUBTREE from a non Q-SUBTREE, and it also makes the description of algorithms very complex. In this chapter, we adopt a different data graph model [Golenberg et al., 2008; Kimelfeld and Sagiv, 2006b], by virtually adding a keyword node for every word w appears in the data and by adding a directed edge from each node v to w with weight 0 if v contains w. Denote the augmented graph as $G_D^A = (V^A, E^A)$. Figure 3.1(e) shows the augmented graph of the graph in the dotted rectangle. Although in Figure 3.1(e), there is only one incoming edge for each keyword node, multiple incoming edges into keyword nodes are allowed in general. Note that, there is only one keyword node for each word w in G_D^A, and the augmented graph does not need to be materialized; it can be built on-the-fly using the inverted index of keywords. In G_D^A, an answer of a keyword query is well defined and captured by the following lemma.

Lemma 3.3   *[Kimelfeld and Sagiv, 2006b] A subtree T of G_D^A is a Q-SUBTREE, for a keyword query $Q = \{k_1, \cdots, k_l\}$, if and only if the set of leaves of T is exactly Q, i.e., $leaves(T) = Q$, and the root of T has at least two children.*

The last three subtrees in Figure 3.2 all satisfy the requirements of Lemma 3.3, so they are Q-SUBTREE. In the following, we also use G_D to denote the augmented graph G_D^A when the context is clear, and we use the above lemma to characterize Q-SUBTREE. Although Q-SUBTREE is popularly used to describe answers to keyword queries, two different weight functions are proposed in the literature to rank Q-SUBTREES in increasing weight order. Two semantics are proposed based on the two weight functions, namely *steiner tree-based semantics* and *distinct root-based semantics*.

Steiner Tree-Based Semantics: In this semantics, the weight of a Q-SUBTREE is defined as the total weight of the edges in the tree; formally,

$$w(T) = \sum_{\langle u,v \rangle \in E(T)} w_e(\langle u, v \rangle) \tag{3.3}$$

where $E(T)$ is the set of edges in T. The l-keyword query finds all (or top-k) Q-SUBTREES in weight increasing order, where the weight denotes the cost to connect the l keywords. Under this semantics, finding the Q-SUBTREE with the smallest weight is the well-known *optimal steiner tree problem*, which is NP-complete [Dreyfus and Wagner, 1972].

Distinct Root-Based Semantics: Since the problem of keyword search under the steiner tree-based semantics is generally a hard problem, many works resort to easier semantics. Under the distinct root-based semantics, the weight of a Q-SUBTREE is the sum of the shortest distance from the root to each keyword node; more precisely,

$$w(T) = \sum_{i=1}^{l} dist(root(T), k_i) \tag{3.4}$$

where $root(T)$ is the root of T, $dist(root(T), k_i)$ is the shortest distance from the root to the keyword node k_i.

There are two differences between the two semantics. First is the weight function as shown above. The other difference is the total number of Q-SUBTREES for a keyword query. In theory, there can be exponentially many Q-SUBTREES under the steiner tree semantics, i.e., $O(2^m)$ where m is the number of edges in G_D. But, under the distinct root semantics, there can be at most n, which is the number of nodes in G_D, Q-SUBTREES, i.e., zero or one Q-SUBTREE rooted at each node $v \in V(G_D)$. The potential Q-SUBTREE rooted at v is the union of the shortest path from v to each keyword node k_i.

3.2   POLYNOMIAL DELAY AND DIJKSTRA'S ALGORITHM

Before we show algorithms to find Q-SUBTREES for a keyword search query, we first discuss two important concepts, namely, *polynomial delay* and θ-approximation, which is used to measure the efficiency of enumeration algorithms, and two algorithms, namely, *Lawler's procedure* for enumerating answers, which is a general procedure to enumerate structural results (e.g., Q-SUBTREE) efficiently, and *Dijkstra's single source shortest path algorithm*, which is a fundamental operation for many algorithms.

Polynomial Delay: For an instance of a problem that consists of an input x and a finite set $\mathcal{A}(x)$ of answers, there is a *weight function* that maps each answer $a \in \mathcal{A}(x)$ to a positive real value, $w(a)$.

An *enumeration algorithm* E is said to enumerate $\mathcal{A}(x)$ in *ranked order* if the output sequence by E, $a_1, \cdots, a_n$, comprises the whole answer set $\mathcal{A}(x)$, and $w(a_i) \leq w(a_j)$ and $a_i \neq a_j$ holds for all $1 \leq i < j \leq n$, i.e., the answers are output in increasing weight order without repetition. For an enumeration algorithm E, there is a delay between outputting two successive answers. There is also a delay before outputting the first answer, or there is a delay after outputting the last result and determining that there are no more answers. More precisely, the i-th delay ($1 \leq i \leq n + 1$) is the length of the time interval that starts immediately after outputting the $(i - 1)$-th answer (or the starting time of the execution of the algorithm if $i - 1 = 0$), and it ends when the i-th answer is output (or the ending time of the execution of the algorithm if no more answer exists). An algorithm E enumerates $\mathcal{A}(x)$ in *polynomial delay* if all the delays can be bounded by polynomial in the size of the input [Johnson et al., 1988]. As a special case, when there is no answer, i.e., $\mathcal{A}(x) = \emptyset$, algorithm E should terminate in time polynomial to the size of input.

There are two kinds of enumeration algorithms with polynomial delay, one enumerates in exact rank order with polynomial delay, the other enumerates in approximate rank order with polynomial delay. In the remainder of this section, we assume that the enumeration algorithm has polynomial delay, so we do not state it explicitly.

θ-**approximation:** Sometimes, enumerating in approximate rank order but with smaller delay is more desirable for efficiency. For an approximation algorithm, the quality is determined by an *approximation ratio* $\theta > 1$ (θ may be a constant, or a function of the input x). A θ-approximation of an optimal answer, over input x, is any answer $app \in \mathcal{A}(x)$, such that $w(app) \leq \theta \cdot w(a)$ for all $a \in \mathcal{A}(x)$. Note that $\perp$ is a θ-approximation if $\mathcal{A}(x) = \emptyset$. An algorithm E enumerates $\mathcal{A}(x)$ in θ-approximation order, if the weight of answer $a_i \in \mathcal{A}(x)$ is at most θ times worse than $a_j \in \mathcal{A}(x)$ for any answer pair (a_i, a_j) where a_i precedes a_j in the output sequence. Typically, the first answer output by E is a θ-approximation of the best answer.

The enumeration algorithms that enumerate all the answers in (θ-approximate) rank order, can find (θ-approximate) top-k answers (or all answers if there are fewer than k answers), by stopping the execution immediately after finding k answers. A θ-approximation of top-k answers is any set $AppTop$ of $\min(k, |\mathcal{A}(x)|)$ answers, such that $w(a) \leq \theta \cdot w(a')$ holds for all $a \in AppTop$ and $a' \in \mathcal{A}(x) \backslash AppTop$ [Fagin et al., 2001]. There are two advantages of enumeration algorithms with polynomial delay to find top-k answers: first, the total running time is linear in k and polynomial in the size of input x; second, k need not be known in advance, the user can decide whether more answers are desired based on the output ones.

Lawler's Procedure: Most of the algorithms that enumerate top-k (or all) answers in polynomial delay is an adaptation of *Lawler's procedure* [Lawler, 1972]. Lawler's procedure generalizes an algorithm of finding top-k shortest path [Yen, 1971] to compute the top-k answers of discrete optimization problems. The main idea of Lawler's procedure is as follows. It considers the whole answer set as an *answer space*. The first answer is the optimal answer in the whole space. Then, the Lawler's procedure works iteratively, and in every iteration, it partitions the subspace (a subset of answers) where the previously output answer comes from, into several subspaces (excluding the previously output

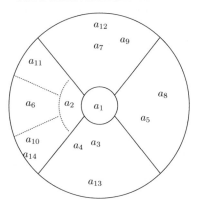

Figure 3.3: Illustration of Lawler's Procedure [Golenberg et al., 2008]

answer) and finds an optimal answer in each newly generated subspace, and the next answer to be output in rank order can be determined to be the optimal among all the answers that have been found but not output.

Example 3.4   Suppose we want to enumerate all the elements in Figure 3.3 in increasing distance from the center, namely, in the order, $a_1, a_2, \cdots, a_{14}$. Initially, the only space consists of all the elements, i.e., $S = \{a_1, a_2 \cdots, a_{14}\}$, and the closest element in S is a_1. In the first iteration, we output a_1 and partition S into 4 subspaces, and the closest element in each subspace is found, namely, a_2 in the subspace $S_1 = \{a_2, a_6, \cdots, a_{14}\}$, a_3 in the subspace $S_2 = \{a_3, a_4, a_{13}\}$, a_5 in the subspace $S_3 = \{a_5, a_8\}$, and a_7 in the subspace $S_4 = \{a_7, a_9, a_{12}\}$. In the second iteration, among all the found but not output elements, i.e., $\{a_2, a_3, a_5, a_7\}$, element a_2 is output to be the next element in rank order, and the subspace S_2 is partitioned into three new subspaces and the optimal element in each subspace is found, i.e., a_{11} in $S_{11} = \{a_{11}\}$, a_6 in $S_{12} = \{a_6\}$, and a_{10} in $S_{13} = \{a_{10}, a_{14}\}$. The next element output is a_3, and the iterations continue.

Dijkstra's Algorithm: Dijkstra's single source shortest path algorithm is designed to find the shortest distance (and the corresponding path) from a source node to every other node in a graph. In the literature of keyword search, the Dijkstra's algorithm is usually implemented as an iterator, and it works on the graph by reversing the direction of every edge. When an iterator is called, it will return the next node that can reach the source with shortest distance among all the unreturned nodes. We will describe an iterator implementation of Dijkstra's algorithm by backward search.

Algorithm 16 (SPITERATOR) shows the two procedures to run Dijkstra's algorithm as an iterator. There are two main data structures, *SPTree* and *Fn*. *SPTree* is a shortest path tree that contains all the explored nodes, which are those nodes whose shortest distance to the source node have been computed. It can be implemented by storing the child of each node v, as $v.pre$. Note

---

Algorithm 16 SPITERATOR (G, s)

Input: a directed graph G, and a source node $s \in V(G)$.

Output: each call of Next returns the next node that can reach s.

1: **Procedure** Initialize()
2:    $SPTree \leftarrow \emptyset; Fn \leftarrow \emptyset$
3:    $s.d \leftarrow 0; s.pre \leftarrow \bot$
4:    $Fn.\text{INSERT}(s)$

5: **Procedure** Next()
6:    **return** $\bot$, if $Fn = \emptyset$
7:    $v \leftarrow Fn.\text{POP}()$
8:    **for each** incoming edge of v, $\langle u, v \rangle \in E(G)$ **do**
9:      **if** $v.d + w_e(\langle u, v \rangle) < u.d$ **then**
10:        $u.d \leftarrow v.d + w_e(\langle u, v \rangle); u.pre \leftarrow v$
11:        $Fn.\text{UPDATE}(u)$ if $u \in Fn$, $Fn.\text{INSERT}(u)$ otherwise
12:    $SPTree.\text{INSERT}(\langle v, v.pre \rangle)$
13:    **return** v

---

that *SPTree* is a reversed tree: every node has only one child but multiple parents. $v.d$ denotes the distance of a path from node v to the source node, and it is ∞, initially. When v is inserted into *SPTree*, it means that its shortest path and shortest distance to the source have been found. *Fn* is a priority queue that stores the *fringe nodes* v sorted on $v.d$, where a fringe node is one whose shortest path to the source is not yet determined but a path has been found. The main operations in *Fn* are, INSERT, POP, TOP, UPDATE, where INSERT (UPDATE) inserts (updates) an entry into (in) *Fn*, TOP returns the entry with the highest priority from *Fn*, and POP additionally pops out that entry from *Fn* after TOP operation. With the implementation of Fibonacci Heap [Cormen et al., 2001], INSERT and UPDATE can be implemented in $O(1)$ amortized time, POP and TOP can be implemented in $O(\log n)$ time where n is the size of the heap.

    SPITERATOR works as follows. It first initializes *SPTree* and *Fn* to be $\emptyset$. The source node s is inserted into *Fn* with $s.d = 0$ and $s.pre = \bot$. When Next is called, if *Fn* is empty, it means that all the nodes that can reach the source node have been output (line 6). Otherwise, it pops the top entry, v, from *Fn* (line 7). It updates the distance of all the incoming neighbors of v whose shortest distance have not been determined (line 8-11). Then, it inserts v into *SPTree* (line 12) and returns v. Given a graph G with n nodes and m edges, the total time of running Next until it returns $\bot$ is $O(m + n \log n)$.

    The concepts of polynomial delay and θ-approximation are used in Lawler's procedure to enumerate answers of a keyword query in (approximate) rank order with polynomial delay. The algorithm of Lawler's procedure is used in Section 3.3.3 and Section 3.5.2. Dijkstra's algorithm is used in Section 3.3.1 and Section 3.5.2.

3.3   STEINER TREE-BASED KEYWORD SEARCH

In this section, we show three categories of algorithms under the steiner tree-based semantics, where the edges are assigned weights as described earlier, and the weight of a tree is the summation of weights of the edges. First is the backward search algorithm, where the first tree returned is an l-approximation of the optimal steiner tree. Second is a dynamic programming approach, which finds the optimal (top-1) steiner tree in time $O(3^l n + 2^l ((l + \log n)n + m))$. Third is enumeration algorithms with polynomial delay.

3.3.1   BACKWARD SEARCH

Bhalotia et al. [2002] enumerate Q-SUBTREEs using a backward search algorithm searching backwards from the nodes that contain keywords. Given a set of l keywords, they first find the set of nodes that contain keywords, S_i, for each keyword term k_i, i.e., S_i is exactly the set of nodes in $V(G_D)$ that contain the keyword term k_i. This step can be accomplished efficiently using an inverted list index. Let $S = \bigcup_{i=1}^{l} S_i$. Then, the backward search algorithm concurrently runs $|S|$ copies of Dijkstra's single source shortest path algorithm, one for each keyword node v in S with node v as the source. The $|S|$ copies of Dijkstra's algorithm run concurrently using iterators (see Algorithm 16). All the Dijkstra's single source shortest path algorithms traverse graph G_D in reverse direction. When an iterator for keyword node v visits a node u, it finds a shortest path from u to the keyword node v. The idea of concurrent backward search is to find a common node from which there exists a shortest path to at least one node in each set S_i. Such paths will define a rooted directed tree with the common node as the root and the corresponding keyword nodes as the leaves.

A high-level pseudocode is shown in Algorithm 17 (BACKWARDSEARCH [Bhalotia et al., 2002]). There are two heaps, $It Heap$ and $Output$, where $It Heap$ stores the $|S|$ copies of iterators of Dijkstra's algorithm, $Out Heap$ is a result buffer that stores the generated but not output results. In every iteration (line 6), the algorithm picks the iterator whose next node to be returned has the smallest distance (line 7). For each node u, a nodelist $u.L_i$ is maintained, which stores all the keyword nodes in S_i whose shortest distance from u has been computed, for each keyword term k_i. $u.L_i \subset S_i$ and is empty initially (line 12). Consider an iterator that starts from a keyword node, say $v \in S_i$, visiting node u. Some other iterators might have already visited node u and the keyword nodes corresponding to those iterators are already stored in $u.L_j$'s. Thus new connection trees rooted at node u and containing node v need to be generated, which is the set of connected trees corresponding to the cross product tuples from $\{\{v\} \times \prod_{j \neq i} u.L_j\}$ (line 13). Those trees whose root has only one child are discarded (line 17), since the directed tree constructed by removing the root node would also have been generated, and they would be a better answer. After generating all connected trees, node v is inserted into list $u.L_i$ (line 14).

Algorithm 17 BackwardSearch (G_D, Q)

Input: a data graph G_D, and an l-keyword query $Q = \{k_1, \cdots, k_l\}$.

Output: Q-subtrees in increasing weight order.

1: Find the sets of nodes containing keywords: $\{S_1, \cdots, S_l\}$, $\mathcal{S} \leftarrow \bigcup_{i=1}^{l} S_i$
2: $ItHeap \leftarrow \emptyset$; $OutHeap \leftarrow \emptyset$
3: **for** each keyword node, $v \in S$ **do**
4:    Create a single source shortest path iterator, I_v, with v as the source node
5:    $ItHeap$.INSERT(I_v), and the priority of I_v is the distance of the next node it will return
6: **while** $ItHeap \neq \emptyset$ **and** more results required **do**
7:    $I_v \leftarrow ItHeap$.POP()
8:    $u \leftarrow I_v$.NEXT()
9:    **if** I_v has more nodes to return **then**
10:       $ItHeap$.INSERT(I_v)
11:    **if** u is not visited before by any iterator **then**
12:       Create $u.L_i$ and set $u.L_i \leftarrow \emptyset$, for $1 \leq i \leq l$
13:    $CP \leftarrow \{v\} \times \prod_{j \neq i} u.L_j$, where $v \in S_i$
14:    Insert v into $u.L_i$
15:    **for** each $tuple \in CP$ **do**
16:       Create $ResultTree$ from $tuple$
17:       **if** root of $ResultTree$ has only one child **then**
18:          continue
19:       **if** $OutHeap$ is full **then**
20:          Output and remove top result from $OutHeap$
21:       Insert $ResultTree$ into $OutHeap$ ordered by its weight
22: output all results in $OutHeap$ in increasing weight order

---

The connected tress generated by BackwardSearch are only approximately sorted in increasing weight order. Generating all the connected trees followed by sorting would increase the computation time and also lead to a greatly increased time to output the first result. A fixed-size heap is maintained as a buffer for the generated connected trees. Newly generated trees are added into the heap without outputting them (line 21). Whenever the heap is full, the top result tree is output and removed (line 20).

Although BackwardSearch is a heuristic algorithm, the first Q-subtree output is an l-approximation of the optimal steiner tree, and the Q-subtrees are generated in increasing height order. The Q-subtrees generated by BackwardSearch is not complete, as BackwardSearch only considers the shortest path from the root of a tree to nodes containing keywords.

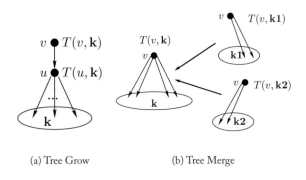

(a) Tree Grow   (b) Tree Merge

Figure 3.4: Optimal Substructure [Ding et al., 2007]

3.3.2   DYNAMIC PROGRAMMING

Although finding the optimal steiner tree (top-1 Q-SUBTREE under the steiner tree-based semantics) or group steiner tree is NP-complete in general, there are efficient algorithms to find the
optimal steiner tree for l-keyword queries [Ding et al., 2007; Kimelfeld and Sagiv, 2006a]. The algorithm [Ding et al., 2007] solves the group steiner tree problem, but the group steiner tree in a
directed (or undirected) graph can be transformed into steiner tree problem in directed graph (the
same as our augmented data graph G_D^A). So, in the following, we deal with the steiner tree problem
(actually, the algorithm is almost the same).

The algorithm is dynamic programming based, whose main idea is illustrated by Figure 3.4.
We use $\mathbf{k}, \mathbf{k1}, \mathbf{k2}$ to denote a non-empty subset of the keyword nodes $\{k_1, \cdots, k_l\}$. Let $T(v, \mathbf{k})$
denote the tree with the minimum weight (called it *optimal tree*) among all the trees, that rooted
at v and containing all the keyword nodes in $\mathbf{k}$. There are two cases: (1) the root node v has only
one child, (2) v has more than one child. If the root node v has only one child u, as shown in
Figure 3.4(a), then the tree $T(u, \mathbf{k})$ must also be an optimal tree rooted at u and containing all
the keyword nodes in $\mathbf{k}$. Otherwise, v has more than one child, as shown in Figure 3.4(b). Assume
the children nodes are $\{u_1, u_2, \cdots, u_n\}(n \leq |\mathbf{k}|)$, and for any partition of the children nodes into
two sets, CH_1 and CH_2, e.g., $CH_1 = \{u_1\}$ and $CH_2 = \{u_2, \cdots, u_n\}$, let $\mathbf{k1}$ and $\mathbf{k2}$ be the set of
keyword nodes that are descendants of CH_1 and CH_2 in $T(v, \mathbf{k})$, respectively. Then $T(v, \mathbf{k1})$ (the
subtree of $T(v, \mathbf{k})$ by removing CH_2 and all the descendants of CH_2), and $T(v, \mathbf{k2})$ (the subtree
of $T(v, \mathbf{k})$ by removing CH_1 and all the descendants of CH_1) must be the corresponding optimal
tree rooted at v and containing all the keyword nodes in $\mathbf{k1}$ and $\mathbf{k2}$, respectively. This means that
$T(v, \mathbf{k})$ satisfies the optimal substructure property, which is needed for the correctness of a dynamic
programming [Cormen et al., 2001].

Based on the above discussions, we can find the optimal tree $T(v, \mathbf{k})$ for each $v \in V(G_D)$
and $\mathbf{k} \subseteq Q$. Initially, for each keyword node k_i, $T(k_i, \{k_i\})$ is a single node tree consisting of the

---

Algorithm 18 DPBF (G_D, Q)

Input: a data graph G_D, and an l-keyword query $Q = \{k_1, k_2, \cdots, k_l\}$.

Output: optimal steiner tree contains all the l keywords.

1: Let $\mathcal{Q}_T$ be a priority queue sorted in the increasing order of weights of trees, initialized to be $\emptyset$
2: **for** $i \leftarrow 1$ **to** l **do**
3:    Initialize $T(k_i, \{k_i\})$ to be a tree with a single node k_i; $\mathcal{Q}_T$.INSERT($T(k_i, \{k_i\})$)
4: **while** $\mathcal{Q}_T \neq \emptyset$ **do**
5:    $T(v, \mathbf{k}) \leftarrow \mathcal{Q}_T$.POP()
6:    **return** $T(v, \mathbf{k})$, if $\mathbf{k} = Q$
7:    **for** each $\langle u, v \rangle \in E(G_D)$ **do**
8:       **if** $w(\langle u, v \rangle \oplus T(v, \mathbf{k})) < w(T(u, \mathbf{k}))$ **then**
9:          $T(u, \mathbf{k}) \leftarrow \langle u, v \rangle \oplus T(v, \mathbf{k})$
10:         $\mathcal{Q}_T$.UPDATE($T(u, \mathbf{k})$)
11:   $\mathbf{k1} \leftarrow \mathbf{k}$
12:   **for** each $\mathbf{k2} \subset Q$, s.t. $\mathbf{k1} \cap \mathbf{k2} = \emptyset$ **do**
13:      **if** $w(T(v, \mathbf{k1}) \oplus T(v, \mathbf{k2})) < w(T(v, \mathbf{k1}) \cup \mathbf{k2})$ **then**
14:         $T(v, \mathbf{k1} \cup \mathbf{k2}) \leftarrow T(v, \mathbf{k1}) \oplus T(v, \mathbf{k2})$
15:         $\mathcal{Q}_T$.UPDATE($T(v, \mathbf{k1} \cup \mathbf{k2})$)

---

keyword node k_i with tree weight 0. For a general case, the $T(v, \mathbf{k})$ can be computed by the following equations.

$$T(v, \mathbf{k}) = \min(T_g(v, \mathbf{k}), T_m(v, \mathbf{k})) \tag{3.5}$$

$$T_g(v, \mathbf{k}) = \min_{\langle v, u \rangle \in E(G_D)} \{\langle v, u \rangle \oplus T(u, \mathbf{k})\} \tag{3.6}$$

$$T_m(g, \mathbf{k1} \cup \mathbf{k2}) = \min_{\mathbf{k1} \cap \mathbf{k2} = \emptyset} \{T(v, \mathbf{k1}) \oplus T(v, \mathbf{k2})\} \tag{3.7}$$

Here, min means to choose the tree with minimum weight from all the trees in the argument. Note that, $T(v, \mathbf{k})$ may not exist for some v and $\mathbf{k}$, which reflects that node v can not reach some of the keyword nodes in $\mathbf{k}$, then $T(v, \mathbf{k}) = \bot$ with weight ∞. $T_g(v, \mathbf{k})$ reflects the case that the root of $T(v, \mathbf{k})$ has only one child, and $T_m(v, \mathbf{k})$ reflects that the root has more than one child.

Algorithm 18 (DPBF, which stands for Best-First Dynamic Programming [Ding et al., 2007]) is a dynamic programming approach to compute the optimal steiner tree that contains all the keyword nodes. Here $T(v, \mathbf{k})$ denotes a tree structure, $w(T(v, \mathbf{k}))$ denotes the weight (see Eq. 3.3) of tree $T(v, \mathbf{k})$, and $T(v, \mathbf{k})$ is initialized to be $\bot$ with weight ∞, for all $v \in V(G_D)$ and $\mathbf{k} \subseteq Q$. DPBF maintains intermediate trees in a priority queue $\mathcal{Q}_T$, by increasing order of the weights of trees. The smallest weight tree is maintained at the top of the queue $\mathcal{Q}_T$. DPBF first initializes $\mathcal{Q}_T$ to be empty (line 1), and inserts $T(k_i, \{k_i\})$ with weight 0 into $\mathcal{Q}_T$ (lines 2-3), for each keyword node in the query, i.e., $\forall k_i \in Q$. While the queue is non-empty and the optimal result has not been found, the algorithm repeatedly updates (or inserts) the intermediate trees $T(v, \mathbf{k})$. It first dequeues the top tree $T(v, \mathbf{k})$ from queue $\mathcal{Q}_T$ (line 5), and this tree $T(v, \mathbf{k})$ is guaranteed to have the smallest weight among all the trees rooted at v and containing the keyword set $\mathbf{k}$. If $\mathbf{k}$ is the whole keyword set, then the algorithm has found the optimal steiner tree that contains all the keywords (line 6). Otherwise, it uses the tree $T(v, \mathbf{k})$ to update other partial trees whose optimal tree structure may contain $T(v, \mathbf{k})$ as a subtree. There are two operations to update trees, namely, Tree Growth (Figure 3.4(a)) and Tree Merge (Figure 3.4(b)). Lines 7-10 correspond to the tree growth operations, and lines 12-15 are the tree merge operations.

Consider a graph G_D with n nodes and m edges, DPBF finds the optimal steiner three containing all the keywords in $Q = \{k_1, \cdots, k_l\}$, in time $O(3^l n + 2^l((l + n) \log n + m))$ [Ding et al., 2007].

DPBF can be modified slightly to output k steiner trees in increasing weight order, denoted as DPBF-K, by terminating DPBF after finding k steiner trees that contain all the keywords (line 6). Actually, if we terminate DPBF when queue $\mathcal{Q}_T$ is empty (i.e., removing line 6), DPBF can find at most n subtrees, i.e., $T(v, Q)$ for $\forall v \in V(G_D)$, where each tree $T(v, Q)$ is an optimal tree among all the trees rooted at v and containing all the keywords. Note that, (1) some of the trees returned by DPBF-K may not be Q-SUBTREE because the root v can have one single child in the returned tree; (2) the trees returned by DPBF-K may not be the true top-k Q-SUBTREEs, namely, the algorithm may miss some Q-SUBTREEs, whose weight is smaller than the largest tree returned.

3.3.3   ENUMERATING Q-SUBTREES WITH POLYNOMIAL DELAY

Although BACKWARDSEARCH can find an l-approximation of the optimal Q-SUBTREE, and DPBF can find the optimal Q-SUBTREE, the non-first results returned by these algorithms can not guarantee their quality (or approximation ratio), and the delay between consecutive results can be very large. In the following, we will show three algorithms to enumerate Q-SUBTREEs in increasing (or θ-approximate increasing) weight order with polynomial delay: (1) an enumeration algorithm enumerates Q-SUBTREEs in increasing *weight* order with polynomial delay under the *data complexity*, (2) an enumeration algorithm enumerates Q-SUBTREEs in $(\theta + 1)$-approximate *weight* order with polynomial delay under *data-and-query complexity*, (3) an enumeration algorithm enumerates Q-SUBTREEs in 2-approximate *height* order with polynomial delay under *data-and-query complexity*.

The algorithms are adaption of the Lawler's procedure to enumerate Q-SUBTREEs in rank order [Golenberg et al., 2008; Kimelfeld and Sagiv, 2006b]. There are two problems that should be solved in order to apply Lawler's procedure: first, how to divide a subspace into subspaces; second, how to find the top-ranked answer in each subspace. First, we discuss a basic framework to address the first problem. Then, we discuss three different algorithms to find the top-ranked answer in each subspace with tree different requirement of the answer, respectively.

Basic Framework: Algorithm 19 (ENUMTREEPD [Golenberg et al., 2008; Kimelfeld and Sagiv, 2006b]) enumerates Q-SUBTREEs in rank order with polynomial delay. In ENUMTREEPD, the space consists of all the answers (i.e., Q-SUBTREEs) of a keyword query Q over data graph G_D. A

Algorithm 19 ENUMTREEPD (G_D, Q)

Input: a data graph G_D, and an l-keyword query $Q = \{k_1, k_2, \cdots, k_l\}$.

Output: enumerate Q-SUBTREEs in rank order.

1: $\mathcal{Q}_T \leftarrow$ an empty priority queue
2: $T \leftarrow$ Q-SUBTREE $(G_D, Q, \emptyset, \emptyset)$
3: **if** $T \neq \bot$ **then**
4:     $\mathcal{Q}_T$.INSERT($\langle\emptyset, \emptyset, T\rangle$)
5: **while** $\mathcal{Q}_T \neq \emptyset$ **do**
6:     $\langle I, E, T\rangle \leftarrow \mathcal{Q}_T$.POP(); ouput(T)
7:     $\langle e_1, \cdots, e_h\rangle \leftarrow$ SERIALIZE $(E(T)\backslash I)$
8:     **for** $i \leftarrow 1$ **to** h **do**
9:         $I_i \leftarrow I \cup \{e_1, \cdots, e_{i-1}\}$
10:         $E_i \leftarrow E \cup \{e_i\}$
11:         $T_i \leftarrow$ Q-SUBTREE (G_D, Q, I_i, E_i)
12:         **if** $T_i \neq \bot$ **then**
13:             $\mathcal{Q}_T$.INSERT($\langle I_i, E_i, T_i\rangle$)

---

subspace is described by a set of *inclusion edges*, I, and a set of *exclusion edges*, E, i.e., it denotes the set of answers, where each of them contains all the edges in I and no edge from E. Intuitively, I and E specify a set of constraints on the answer of query Q over G_D, where *inclusion edges* specifies that each answer should contain all the edges in I, and *exclusion edges* specifies that each answer should not include any edges from E. We use pair $\langle I, E\rangle$ to denote a subspace. The algorithm uses a priority queue $\mathcal{Q}_T$. An element in $\mathcal{Q}_T$ is a triplet $\langle I, E, T\rangle$, where $\langle I, E\rangle$ describes a subspace and T is the tree found by algorithm Q-SUBTREE from that subspace. Priority of $\langle I, E, T\rangle$ in $\mathcal{Q}_T$ is based on the weight (or height) of T.

    ENUMTREEPD starts by finding a best tree T in the whole space, i.e., space $\langle\emptyset, \emptyset\rangle$. If $T = \bot$, then there is no answer satisfying the keywords requirement, otherwise, $\langle\emptyset, \emptyset, T\rangle$ is inserted into $\mathcal{Q}_T$. In the main loop of line 5, the top ranked triplet $\langle I, E, T\rangle$ is removed from $\mathcal{Q}_T$ (line 6), and T is output as the next Q-SUBTREE in order. $\langle e_1, \cdots, e_h\rangle$ is the sequence of edges of T that are not in I, after serialization by SERIALIZE (which will be discussed shortly) to make the subspaces generated next satisfy some specific property. Next, in lines 8–13, h subspaces $\langle I_i, E_i\rangle$ are generated and T_i, the tree found by Q-SUBTREE in that subspace is found. It is easy to check that, all the subspaces, consisting of the subspaces in $\mathcal{Q}_T$ and the subspaces (each T is also a subspace) that have been output, disjointly comprise of the whole space.

    ENUMTREEPD enumerates all Q-SUBTREEs of G_D. The delay and the order of enumeration are determined by the implementation of Q-SUBTREE (). The following theorem shows that ENUMTREEPD enumerates Q-SUBTREEs in rank order, provided that Q-SUBTREE () returns optimal answers, or in θ-approximate order, provided that Q-SUBTREE () returns θ-approximate answer.

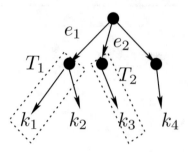

Figure 3.5: SERIALIZE

The theorem also specifies the delay in terms of the running time of Q-SUBTREE (). Recall that n and m are the number of nodes and edges of G_D respectively, l is the number of keywords, and n_i is the number of nodes in the i-th output tree. Note that there are at most 2^m trees, i.e., $i \leq 2^m$.

Theorem 3.5   *[Kimelfeld and Sagiv, 2006b] Consider a data graph G_D and a query $Q = \{k_1, \cdots, k_l\}$.*

- ENUMTREEPD *enumerates all the* Q-SUBTREES *of G_D in the rank order if* Q-SUBTREE *() returns an optimal tree.*

- *If* Q-SUBTREE *() returns a θ-approximation of optimal tree, then* ENUMTREEPD *enumerates in a θ-approximate ranked order.*

- *If* Q-SUBTREE *() terminates in time $t(n, m, l)$, then* ENUMTREEPD *outputs the $(i + 1)$-th answer with delay $O(n_i(t(n, m, l) + \log(n \cdot i) + n_i))$.*

The task of enumerating Q-SUBTREEs is transformed into finding an optimal Q-SUBTREE under a set of constraints, which are specified as inclusion edges I and exclusion edges E. The constraints specified by exclusion edges can be handled easily by removing those edges in E from the data graph G_D. So, in the following, we only consider the inclusion edges I, recall that it is the set of edges that each answer in the subspace should contain. A *partial tree* (PT) is any directed subtree of G_D. A set of PTs $\mathcal{P}$ is called a set of PT constraints if the PTs in $\mathcal{P}$ are pairwise node-disjoint. The set of leaves in the PTs of $\mathcal{P}$ is denoted as $leaves(\mathcal{P})$.

Proposition 3.6   *[Kimelfeld and Sagiv, 2006b] The algorithm* Q-SUBTREE *() can be executed efficiently so that, for every generated set of inclusion edges I, the subgraph of G_D induced by I forms a set of PT constrains $\mathcal{P}$, such that $leaves(\mathcal{P}) \subseteq \{k_1, \cdots, k_l\}$ and $\mathcal{P}$ has at most two PTs.*

SERIALIZE function at line 7 of ENUMTREEPD is used to order the set of edges, such that the newly generated inclusion edges satisfy the above proposition, i.e., $|\mathcal{P}| \leq 2$. The general idea

---

Algorithm 20 SuperTree $(G, \mathcal{T})$

Input: a data graph G, and a set of PT constraints $\mathcal{T}$.
Output: a minimum weight super tree of $\mathcal{T}$ in G.

1: $G' \leftarrow collapse(G, \mathcal{T})$
2: $\mathcal{R} \leftarrow \{root(T) | T \in \mathcal{T}\}$
3: $T' \leftarrow$ SteinerTree $(G', \mathcal{R})$
4: **if** $T' \neq \perp$ **then**
5:      **return** $restore(G, T', \mathcal{T})$
6: **else**
7:      **return** $\perp$

---

of Serialize is shown in Figure 3.5. Assume the tree in Figure 3.5 is T that was obtained in line 6 of EnumTreePD. We regard the problem as recursively adding edges from $E(T) \backslash I$ into $\mathcal{P}$. We discuss two different cases: $|\mathcal{P}| = 1$ and $|\mathcal{P}| = 2$. If $|\mathcal{P}| = 1$, i.e., $\mathcal{P} = \{T_1\}$, then there are two choices, either adding the incoming edge to the root of T_1, e.g., edge e_1, or adding the incoming edge to a keyword node that is not in $V(T_1)$, e.g. the incoming edge to k_2 or k_4. In the other case, $\mathcal{P} = \{T_1, T_2\}$, there are also two choices: either adding the incoming edge to the root of T_1, e.g., edge e_1, or adding the incoming edge to the root of T_2, e.g., edge e_2, and, eventually, T_1 and T_2 will be merged into one tree.

In $\mathcal{P}$, there are two types of PTs. A *reduced* PT (RPT) has a root with at least two children, whereas a *nonreduced* PT (NPT) has a root with only one child. As a special case, a single node is considered as an RPT. Without loss of generality, it can add to $\mathcal{P}$ every keyword node not appearing in $leaves(\mathcal{P})$ as a single node RPT with that keyword node. Thus, from now on, we assume that $leaves(\mathcal{P}) = \{k_1, \cdots, k_l\}$, and there can be more than two PTs, but $\mathcal{P}$ can have at most two NPTs and also at most two non-single node PTs. We denote $\mathcal{P}_{|RPT}$ and $\mathcal{P}_{|NPT}$ as the set of all the RPTs and the set of all the NPTs of $\mathcal{P}$, respectively.

In the following, we discuss different implementations of Q-subtree (G_D, Q, I, E). We first create another graph G by removing those edges in E from G_D, and I forms a set of PT constraints as described above. So we assume that the inputs of the algorithm are a data graph G and a set of PT constraints $\mathcal{P}$ where $leaves(\mathcal{P}) = Q$.

Finding Minimum Weight Super Tree of $\mathcal{P}$: We first introduce a procedure to find a minimum weight super tree of $\mathcal{P}$, i.e., a tree T that contains $\mathcal{P}$ as subtrees. Sometimes, the found super tree is also an optimal Q-subtree, but it may not be reduced. For example, for the two PTs, T_1 and T_2 in the upper left part of Fig. 3.6, the tree with T_1 and T_2 and the edge $\langle v_2, v_5 \rangle$ is a minimum weight super tree, but it is not reduced, so it is not a Q-subtree.

Algorithm 20 (SuperTree [Kimelfeld and Sagiv, 2006b]) finds the optimal super tree of $\mathcal{T}$ if it exists. It reduces the problem to a steiner tree problem by collapsing graph G according to $\mathcal{T}$.

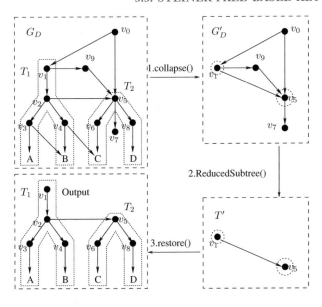

Figure 3.6: Execution example of finding supertree [Kimelfeld and Sagiv, 2006b]

The graph $collapse(G, \mathcal{T})$ is the result of collapsing all the subtrees in $\mathcal{T}$, and it can be obtained as follows.

- Delete all the edges $\langle u, v \rangle$, where v is a non-root node of a PT $T \in \mathcal{T}$ and $\langle u, v \rangle$ is not an edge of T.

- For the remaining edges $\langle u, v \rangle$, such that u is a non-root node of a PT $T \in \mathcal{T}$ and $\langle u, v \rangle$ is not an edge of T, add an edge $\langle root(T), v \rangle$. The weight of the edges $\langle root(T), v \rangle$ is the minimum among the weights of all such edges (including the original edges in G).

- Delete all the non-root nodes of PTs of $\mathcal{T}$ and their associated edges.

As an example, the top part of Figure 3.6 shows how two node-disjoint subtrees T_1 and T_2 are collapsed. In this figure, the edge weights are not shown, and they are assumed equal. In the collapsed graph G', it needs to find a minimum directed steiner tree to contain all the root nodes of the PTs in $\mathcal{T}$ (line 3), this step can be accomplished by existing algorithms. Next, it needs to restore T' to be a super tree of $\mathcal{T}$ in G. First, it adds back all the edges of each PT $T \in \mathcal{T}$ to T'. Then, it replaces each edge in T' with the original edge from which the collapse step gets (it can be the edge itself).

Figure 3.6 shows the execution of SUPERTREE for the input consisting of G and $\mathcal{T} = \{T_1, T_2\}$. In the first step, G' is obtained from G by collapsing the subtrees T_1 and T_2. The second step constructs

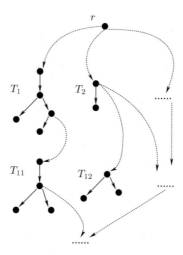

Figure 3.7: The high-level structure of a reduced minimum steiner tree

a minimum directed steiner tree T' of G' with respect to the set of roots $\{v_1, v_5\}$. Finally, T_1 and T_2 are restored in T' and the result is returned.

Theorem 3.7   *[Kimelfeld and Sagiv, 2006b] Consider a data graph G_D and a set $\mathcal{T}$ of PT constraints. Let n and m be the number of nodes and edges of G_D respectively, and let t be number of PTs in $\mathcal{T}$.*

- MinWeightSuperTree, *in which the* SteinerTree *is implemented by* DPBF, *returns a minimum weight super tree of $\mathcal{T}$ if one exists, or $\perp$ otherwise. The running time is $O(3^t n + 2^t((l + n)\log n + m))$.*

- AppWeightSuperTree, *in which the* SteinerTree *is implemented by a $\theta(n, m, t)$-approximation algorithm with running time $f(n, m, t)$, returns a $\theta(n, m, t)$-approximate minimum weight super tree of $\mathcal{T}$ if one exists, or $\perp$ otherwise. The running time is $O(m \cdot t + f(n, m, t))$.*

Finding minimum weight Q-subtree **under** $\mathcal{P}$**:** The minimum weight super tree of $\mathcal{P}$ returned by MinWeightSuperTree is sometimes a Q-subtree, but it is not reduced other times. This situation is caused by the fact that some PTs in $\mathcal{P}$ are NPTs, and the root of one of these NPTs becomes the root of the tree returned by MinWeightSuperTree. So, if it can find the true root of the minimum weight Q-subtree, then it can find the answer by MinWeightSuperTree. Now let's analyze a general minimum weight Q-subtree as shown in Figure 3.7, where $T_1, \cdots, T_{11}, \cdots$ are PTs of $\mathcal{P}$, solid arrows denote paths in a PT, and a dotted arrow denotes a path with no node from $\mathcal{P}$ except the start and end nodes. Node r is the root node, and it can be a root node from $\mathcal{P}$. For each PT $T \in \mathcal{P}$, there can be at most one incoming edge to $root(T)$ and no incoming edges to non-root nodes of T. Let *level* for every $T \in \mathcal{P}$ be $level(T)$, which is the number of different PTs on the path from root to this PT. For example, $level(T_1) = level(T_2) = 0$ and $level(T_{11}) = level(T_{12}) = 1$. We only care about the PTs at level 0, which we call top-level PTs, and denoted them as $\mathcal{T}_{top}$. First, assume $|\mathcal{T}_{top}| \geq 2$. We use T_{top} to denote the subtree consisting of all the paths from r to the root node of PTs in $\mathcal{T}_{top}$ and their associated nodes. We denote the union of T_{top} and $\mathcal{T}_{top}$ as T_{top}^+, i.e., $T_{top}^+ = T_{top} \cup \mathcal{T}_{top}$. The case, $|\mathcal{T}_{top}| = 1$, is implicitly captured by the cases of $|\mathcal{T}_{top}| \geq 2$. Note that, T_{top} may not be reduced, i.e., the root may have only one child, but T_{top}^+ will be a reduced tree.

The algorithm to find a minimum weight Q-SUBTREE under PT constraints $\mathcal{P}$ consists of three steps. First, we assume that, the set of top-level PTs, $\mathcal{T}_{top}$ is found.

1. Find a minimum weight super tree T_{top} in G^1 with the set of root nodes in $\mathcal{T}_{top}$ as the terminal nodes, where G^1 is obtained from G by deleting all the nodes in $\mathcal{P}$ except those root nodes in $\mathcal{T}_{top}$. It is easy to verify that, T_{top} can be found this way.

2. Union T_{top} and $\mathcal{T}_{top}$ to get the expanded tree T_{top}^+.

3. Find a minimum weight super tree of $\mathcal{P} \backslash \mathcal{T}_{top} \cup \{T_{top}^+\}$ from G^2, where G^2 is obtained by deleting all the incoming edges to $root(T_{top}^+)$. This step is to ensure that $root(T_{top}^+)$ will be the root of the final tree.

The above steps can find a minimum weight Q-SUBTREE under constraints $\mathcal{P}$, given $\mathcal{T}_{top}$. Usually, it is not easy to find $\mathcal{T}_{top}$. However, it can resort to an exponential time algorithm that enumerates all the subsets of $\mathcal{P}$ and finds an optimal Q-SUBTREE with each of the subsets as $\mathcal{T}_{top}$. The tree with minimum weight will be the final Q-SUBTREE under constraints $\mathcal{P}$.

The detailed algorithm is shown in Algorithm 21 (MINWEIGHTQSUB-TREE [Kimelfeld and Sagiv, 2006b]). It handles two special cases in lines 1-2 where $\mathcal{P}$ contains only one PT. The non-root nodes of NPTs in $\mathcal{P}$ are removed to avoid finding a non-reduced tree (line 3). Then it enumerates all the possible the top-level PTs (line 5). For each possible top-level PTs, $\mathcal{T}_{top}$, it first finds T_{top} by calling MINWEIGHTSUPERTREE (line 7), then gets T_{top}^+ (line 9), and finds a minimum weight Q-SUBTREE with $root(T_{top}^+)$ as the root (lines 10-11). Note that, data graph G is not necessarily generated as G_D^A for a keyword search problem; MINWEIGHTQSUBTREE works for any general directed graph, i.e., the terminal nodes can also have outgoing edges.

Theorem 3.8 *[Kimelfeld and Sagiv, 2006b] Consider a data graph G with n nodes and m edges. Let $Q = \{k_1, \cdots, k_l\}$ be a keyword query and $\mathcal{P}$ be a set of p PT constraints, such that $leaves(\mathcal{P}) = Q$. MINWEIGHTQSUBTREE returns either a minimum weight Q-SUBTREE containing $\mathcal{P}$ if one exists, or $\perp$ otherwise. The running time of MINWEIGHTQSUBTREE is $O(4^p + 3^p((l + \log n)n + m))$.*

Finding $(\theta + 1)$-approximate minimum weight Q-SUBTREE under $\mathcal{P}$: In this part, we assume that APPWEIGHTSUPERTREE can find a θ-approximation of minimum steiner tree in polynomial time $f(n, m, t)$. Then, MINWEIGHTQSUBTREE can be modified to find a θ-approximation of the minimum weight Q-SUBTREE, by replacing MINWEIGHTSUPERTREE with APPWEIGHTSUPERTREE.

Algorithm 21 MINWEIGHTQSUBTREE $(G, \mathcal{P})$

Input: a data graph G, and a set of PT constraints $\mathcal{P}$.
Output: a minimum weight Q-SUBTREE under constraints $\mathcal{P}$.

1: **return** T, if $\mathcal{P}$ consists of a single RPT T
2: **return** $\bot$, if $\mathcal{P}$ consists of a single NPT
3: let $\mathcal{T}$ be obtained from $\mathcal{P}$ by replacing each $N \in \mathcal{P}_{|NPT}$ with the PT consisting of only the node $root(N)$
4: Initialize $T_{min} \leftarrow \bot$, where $\bot$ has a weight ∞
5: **for all** subsets $\mathcal{T}_{top}$ of $\mathcal{T}$, such that $|\mathcal{T}_{top}| \geq 2$ **do**
6:     $G^1 \leftarrow G - \cup_{T \in \mathcal{P}} V(T) + \cup_{T \in \mathcal{T}_{top}} \{root(T)\}$
7:     $T_{top} \leftarrow$ MINWEIGHTSUPERTREE $(G^1, \mathcal{T}_{top})$
8:     **if** $T_{top} \neq \bot$ **then**
9:         $T_{top}^+ \leftarrow T_{top} \cup \mathcal{T}_{top}$
10:        let G^2 be obtained from G by removing all the incoming edges to $root(T_{top}^+)$
11:        $T \leftarrow$ MINWEIGHTSUPERTREE $(G^2, \mathcal{P} \backslash \mathcal{T}_{top} \cup \{T_{top}^+\})$
12:        $T_{min} \leftarrow \min\{T_{min}, T\}$
13: **return** T_{min}

But, this will take exponential time, as line 5 in MINWEIGHTQSUBTREE. In the following, we introduce approaches to find a $(\theta + 1)$-approximate minimum weight Q-SUBTREE under constraints $\mathcal{P}$.

Algorithm 22 (APPWEIGHTQSUBTREE [Kimelfeld and Sagiv, 2006b]) finds a $(\theta + 1)$-approximation of the minimum weight Q-SUBTREE in polynomial time under data-and-query complexity. Recall that the number of NPTs of $\mathcal{P}$ is no more than 2. APPWEIGHTQSUBTREE mainly consists of three steps.

1. A minimum weight Q-SUBTREE is found by calling MINWEIGHTQSUBTREE if there are no RPTs in $\mathcal{P}$ (line 1). Otherwise, it finds a θ-approximation of minimum weight super tree of $\mathcal{P}$, T_{app}, by calling APPWEIGHTSUPERTREE (line 2). Return T_{app} if it is a Q-SUBTREE or $\bot$ (line 3).

2. Find a reduced tree T_{min} (lines 4-5). The weight of T_{min} is guaranteed to be smaller than the minimum weight Q-SUBTREE under constraints $\mathcal{P}$ if one exists. Note that this step can be accomplished by a single call to procedure MINWEIGHTQSUBTREE, by adding to G' a virtual node v and an edge to each $root(T)$ for all $T \in \mathcal{P}_{|RPT}$, and calling MINWEIGHTQSUBTREE $(G', \mathcal{P}_{|NPT} \cup \{v\})$. If T_{min} is $\bot$, then there is no Q-SUBTREE satisfying the constraints $\mathcal{P}$ (line 6).

Algorithm 22 AppWeightQSubtree $(G, \mathcal{P})$

Input: a data graph G, and a set of PT constraints $\mathcal{P}$.
Output: a $(\theta + 1)$-approximation of minimum weight Q-subtree under constraints $\mathcal{P}$.

1: **return** MinWeightQSubtree $(G, \mathcal{P})$, if $\mathcal{P}_{|RPT} = \emptyset$
2: $T_{app} \leftarrow$ AppWeightSuperTree $(G, \mathcal{P})$
3: **return** T_{app}, if $T_{app} = \perp$ or T_{app} is reduced
4: let G' be obtained from G by removing all the edges $\langle u, v \rangle$, where v is a non-root node of some $T \in \mathcal{P}$ and $\langle u, v \rangle$ is not an edge in T
5: $T_{min} = \min_{T \in \mathcal{P}_{|RPT}}$ MinWeightQSubtree $(G', \mathcal{P}_{|NPT} \cup \{T\})$
6: **return** $\perp$, if $T_{min} = \perp$
7: $r \leftarrow root(T_{min})$
8: **if** r belongs to a subtree $T \in \mathcal{P}_{|RPT}$ **then**
9: $\quad r \leftarrow root(T)$
10: $G_{app} \leftarrow T_{min} \cup T_{app}$
11: remove from G_{app} all incoming edges of r
12: **for all** $v \in V(G_{app})$ that have two incoming edges $e_1 \in E(T_{min})$ and $e_2 \in E(T_{app})$ **do**
13: $\quad$ remove e_2 from G_{app}
14: delete from G_{app} all structural nodes v, such that no keyword is reachable from v
15: **return** G_{app}

---

3. Union T_{app} and T_{min}, and remove redundant nodes and edges to get an $(\theta + 1)$-approximate Q-subtree (line 7-14). Note that, all the edges in T_{min} are kept during the removal of redundant nodes and edges.

The general idea of AppWeightQSubtree is that it first finds a θ-approximate super tree of $\mathcal{P}$, denoted T_{app}. If T_{app} does not exist, then there is no Q-subtree under constraints $\mathcal{P}$. If T_{app} is reduced, then it is a θ-approximation of the minimum weight Q-subtree. Otherwise, it finds another subtree T_{min}, which is guaranteed to be reduced. If there is a Q-subtree under constraints $\mathcal{P}$, then T_{min} must exist and its weight must be smaller than the minimum Q-subtree because a subtree of the minimum Q-subtree satisfies Line 5. Let r denote the root of T_{min}; there are three cases: either r is the root of a NPT in $\mathcal{P}$, or r is the root node of a RPT in $\mathcal{P}$, or r is not in $\mathcal{P}$. If r is the root of a NPT in $\mathcal{P}$, then it must have at least two children, otherwise the root of T_{app} must have an incoming edge (guaranteed by Line 5), as it is the root of one NPT in $\mathcal{P}$. If both T_{app} and T_{min} exist, then from lines 7-14, it can get a Q-subtree. Since each returned node, except the root node, has exactly one incoming edge, it will form a tree.

Theorem 3.9 *[Kimelfeld and Sagiv, 2006b] Consider a data graph G with n nodes and m edges. Let $Q = \{k_1, \cdots, k_l\}$ be a keyword query and $\mathcal{P}$ be a set of PT constraints, such that leaves$(\mathcal{P}) = Q$ and $\mathcal{P}$ has at most c NPTs. AppWeightQSubtree finds a $(\theta + 1)$-approximation of the minimum weight*

Q-SUBTREE *that satisfies* $\mathcal{P}$ *in time* $O(f + 4^{c+1}n + 3^{c+1}((l + \log n)n + m))$, *where* θ *and* f *are the approximation ratio and runtime, respectively, of* APPWEIGHTSUPERTREE.

Finding 2-approximate minimum height Q-SUBTREE **under** $\mathcal{P}$: Although MINWEIGHTQSUB-TREE and APPWEIGHTQSUBTREE can enumerate Q-SUBTREES in exact (or approximate) rank order, they are based on repeated computations of steiner trees (or approximate steiner trees) under inclusion and exclusion constraints; therefore, they are not practical. Golenberg et al. [2008] propose to decouple Q-SUBTREE ranking step from Q-SUBTREE generation. More precisely, it first generates a set of N Q-SUBTREES that are candidate answers, by incorporating a much easier rank function than the steiner tree weight function, and then generates a set of k final answers which are ranked according to a more complex ranking function. The ranking function used is based on the height, where the height of a tree is the maximum among the shortest distances to each keyword node, i.e., $height(T) = \max_{i=1}^{l} dist(root(T), k_i)$. Ranking in increasing height order is very correlated to the desired ranking [Golenberg et al., 2008], so an enumeration algorithm is proposed to generate Q-SUBTREES in 2-approximation rank order with respect to the height ranking function.

The general idea is the same as that of enumerating Q-SUBTREES in $(\theta + 1)$-approximate order with respect to the weight function, i.e. APPWEIGHTQSUBTREE. It also uses ENUMTREEPD as the outer enumeration algorithm, and it implements the sub-routine Q-SUBTREE () by returning a 2-approximation of minimum height Q-SUBTREE under constraints $\mathcal{P}$. Finding an approximate tree with respect to height ranking function under constraints $\mathcal{P}$ is much easier than with the weight ranking function, i.e., APPWEIGHTQSUBTREE. It also consists of three steps: (1) find a minimum height super tree of $\mathcal{P}$, T_{sup}, and return T_{sup} if it is reduced or equal to $\bot$, (2) otherwise, find another reduced subtree T_{min} whose height is guaranteed to be no larger than that of the minimum height Q-SUBTREE if one exists, (3) return the union of T_{sup} and T_{min} after removing redundant edges and nodes.

The algorithm to find a minimum height super tree of $\mathcal{T}$ is shown in Algorithm 23 (MIN-HEIGHTSUPERTREE [Golenberg et al., 2008]). The general idea is the same as BACKWARDSEARCH, by creating an iterator for each leaf node of $\mathcal{T}$ (lines 3-5). During each execution of Line 6, it first finds the iterator I_v, whose next node to be returned is the one with the shortest distance to its source. Let u to be that node. If u has been returned from all the other iterators (line 9), it means that the shortest paths from u to all the leaf nodes in $\mathcal{T}$ have been computed. The union of these shortest paths is a tree with minimum height to include all the leaf nodes of $\mathcal{T}$. But there is one problem that remains to be solved: the tree returned must be a super tree of $\mathcal{T}$. All the edges $\langle u, v \rangle$ from G, where v is a non-root node of some $T \in \mathcal{T}$ and $\langle u, v \rangle$ is not an edge in T (line 1), can be removed since in a tree every node can have at most one incoming edge and $\langle u, v \rangle$ must be included. This operation makes sure that for every non-root of $\mathcal{T}$, the incoming edge in $\mathcal{T}$ is included. Also, the root of the tree returned can not be a non-root of $\mathcal{T}$, which can be checked by Line 9. Then, the tree returned by MINHEIGHTSUPERTREE is a minimum height super tree of $\mathcal{T}$ in G.

Algorithm 23 MINHEIGHTSUPERTREE $(G, \mathcal{T})$

Input: a data graph G, and a set of PT constraints $\mathcal{T}$.
Output: a minimum height super tree of $\mathcal{T}$ in G.

1: remove all the edges $\langle u, v \rangle$ from G, where v is a non-root node of some $T \in \mathcal{T}$ and $\langle u, v \rangle$ is not an edge in T
2: $ItHeap \leftarrow \emptyset$
3: **for each** leave node, $v \in leaves(\mathcal{T})$ **do**
4:     Create a single source shortest path iterator, I_v, with v as the source node
5:     $ItHeap.\textsc{insert}(I_v)$, the priority of I_v is the distance of the next node it will return
6: **while** $ItHeap \neq \emptyset$ **do**
7:     $I_v \leftarrow ItHeap.\textsc{pop}()$
8:     $u \leftarrow I_v.\textsc{next}()$
9:     **if** u has been returned from all the other iterators **and** u is not a non-root node of any tree $T \in \mathcal{T}$ **then**
10:         **return** the subtree which is the union of the shortest paths from u to each leaf node of $\mathcal{T}$
11:     **if** I_v has more nodes to return **then**
12:         $ItHeap.\textsc{insert}(I_v)$
13: **return** $\perp$

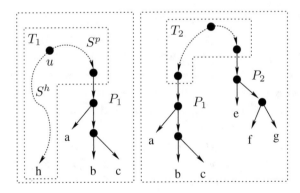

Figure 3.8: Approximating a minimum height Q-SUBTREE [Golenberg et al., 2008]

If the tree T_{app} found by MINHEIGHTSUPERTREE is non-reduced, then it needs to find another reduced tree T_{min}, and the root of T_{app} must be the root of one NPT in $\mathcal{P}$; without loss of generality, we assume it to be P_1. Note that, with respect to the height ranking function, it can not use the idea of MINWEIGHTQSUBTREE to find minimum height Q-SUBTREE. There are two cases to consider depending on whether the following holds: in a minimum height Q-SUBTREE A_m that satisfies $\mathcal{P}$, there is a path from the root to a single node PT in $\mathcal{P}$ that does not use any edge of $\mathcal{P}$.

Algorithm 24 AppHeightQSubtree (G_D, $\mathcal{P}$)

Input: a data graph G, and a set of PT constraints $\mathcal{P}$.

Output: a 2-approximation of minimum height Q-subtree under constraints $\mathcal{P}$.

 1: **return** $\bot$, **if** $\mathcal{P}$ consists of a single NPT
 2: $T_{app} \leftarrow$ MinHeightSuperTree (G, $\mathcal{P}$)
 3: **return** T, **if** $T = \bot$ or T is a Q-subtree
 4: **if** there exists single node RPTs in $\mathcal{P}$ **then**
 5:    construct the tree T_1
 6: **if** there exists two non-single node PTs in $\mathcal{P}$ **then**
 7:    construct the tree T_2
 8: **return** $\bot$, if $T_1 = \bot$ and $T_2 = \bot$
 9: $T_{min} \leftarrow$ minimum height subtree among T_1 and T_2
10: construct a Q-subtree T from T_{app} and T_{min}
11: **return** T

The two cases are shown in Figure 3.8. Essentially, A_m must contain a subtree that looks like either T_1 or T_2. We discuss these cases below.

T_1 describes the following situation: (1) one single node PT (e.g., keyword node h) is reachable from the root of A_m through a path S_h that does not use any edge of $\mathcal{P}$; and (2) P_1 is reachable from the root of A_m through a path S_p that does not include any edge appearing on S_h. Let G^v denote the graph obtained from G by deleting all the non-root nodes of PTs in $\mathcal{P}$, and G^e denote the graph obtained from G by deleting all edges $\langle u, v \rangle$ where v is a non-root node of a PT in $\mathcal{P}$ and $\langle u, v \rangle$ is not in $\mathcal{P}$. For each single node PT in $\mathcal{P}$, e.g., the keyword node h, it can find the minimum height subtree T_h by concurrently running two iterators of Dijkstra's algorithm, one with h as the source and works on G^v, the other with $root(P_1)$ as the source and works on G^e. T_1 is the minimum height subtree among all the found subtrees.

T_2 applies only when $\mathcal{P}$ contains two non-single PTs, P_1 and P_2, where P_2 can be either NPT or RPT. If P_2 is a NPT, then T_2 can not use any edge from $\mathcal{P}$, so it can be found in the graph G^v. Otherwise, P_2 is a RPT, the root of T_2 can be the root of P_2. Then it needs to build a new graph G' from G as follows: (1) remove all the edges entering into non-root nodes of P_2 and are not in P_2 itself (i.e., it is handled as in the construction of G^e); (2) remove all the non-root nodes of P_1 (i.e., it is handled as in the construction of G^v). In G', T_2 can be found by two iterators using Dijkstra's algorithm.

Theorem 3.10   *[Golenberg et al., 2008] Given a data graph G with n nodes and m edges, let $Q = \{k_1, \cdots, k_l\}$ be a keyword query and $\mathcal{P}$ be a set of PT constraints, such that $leaves(\mathcal{P}) = Q$ and $\mathcal{P}$ has at most two non-single node PTs. AppHeightQSubtree finds a 2-approximation of the minimum height Q-subtree that satisfies $\mathcal{P}$ in time $O(l(n \log n + m))$.*

3.4   DISTINCT ROOT-BASED KEYWORD SEARCH

In this section, we show approaches to find Q-SUBTREES using the distinct root semantics, where the weight of a tree is defined as the sum of the shortest distance from the root to each keyword node. As shown in the previous section, the problem of keyword search under the directed steiner tree is, in general, a hard problem. Using the distinct root semantics, there can be at most n Q-SUBTREES for a keyword query, and in the worst case, all the Q-SUBTREES can be found in time $O(l(n \log n + m))$. The approaches introduced in this section deal with very large graphs in general, and they propose search strategies or indexing schemes to reduce the search time for an online keyword query.

3.4.1   BIDIRECTIONAL SEARCH

BACKWARDSEARCH algorithm, as proposed in the previous section, can be directly applied to the distinct root semantics, by modifying Line 3 to iterate over the l keyword nodes, i.e., $\{k_1, \cdots, k_l\}$. It would explore an unnecessarily large number of nodes in the following scenarios:

- The query contains a frequently occurring keyword. In BACKWARDSEARCH, one iterator is associated with each keyword node. The algorithm would generate a large number of iterators if a keyword matches a large number of nodes.

- An iterator reaches a node with large fan-in (incoming edges). An iterator may need to explore a large number of nodes if it hits a node with a very large fan-in.

Bidirectional search [Kacholia et al., 2005] can be used to overcome the drawbacks of BACKWARD-SEARCH. The main idea of bidirectional search is to start forward searches from potential roots. The main difference of bidirectional search from BACKWARDSEARCH are as follows:

- All the single source shortest path iterators from the BACKWARDSEARCH algorithm are merged into a single iterator, called the *incoming iterator*.

- An *outgoing iterator* runs concurrently, which follows forwarding edges starting from all the nodes explored by the *incoming iterator*.

- A *spreading activation* is proposed to prioritize the search, which chooses *incoming iterator* or *outgoing iterator* to be called next. It also chooses the next node to be visited in the *incoming iterator* or *outgoing iterator*.

    A high-level pseudocode for the bidirectional search algorithm is shown in Algorithm 25 (BIDIRECTIONALSEARCH). Q_{in} (Q_{out}) is a priority queue of nodes in backward (forward) expanding fringe. X_{in} (X_{out}) is a set of nodes expanded for incoming (outgoing) paths. Fringe nodes are the set of nodes that have already been hit by an iterator with the neighbors being yet to be explored. The set of fringe nodes of an iterator are called *iterator frontier*. For every node u explored so far, either in outgoing or incoming search, the algorithm keeps track of the best known path from u to any node in S_i. Specifically, for every keyword term k_i, it maintains the child node $sp_{u,i}$ that u follows

---

Algorithm 25 BIDIRECTIONALSEARCH (G_D, Q)

Input: a data graph G_D, and an l-keyword query $Q = \{k_1, k_2, \cdots, k_l\}$.
Output: Q-SUBTREEs in approximately increasing weight order.

1: Find the keyword node sets: $\{S_1, \cdots, S_l\}, \mathcal{S} \leftarrow \bigcup_{i=1}^{l} S_i$
2: $\mathcal{Q}_{in} \leftarrow \mathcal{S}; \mathcal{Q}_{out} \leftarrow \emptyset; X_{in} \leftarrow \emptyset; X_{out} \leftarrow \emptyset$
3: $\forall v \in \mathcal{S} : P_u \leftarrow \emptyset, depth_u \leftarrow 0; \forall i, \forall u \in \mathcal{S} : sp_{u,i} \leftarrow \infty$
4: $\forall i, \forall u \in \mathcal{S} :$ if $u \in S_i, dist_{u,i} \leftarrow 0$ else $dist_{u,i} \leftarrow \infty$
5: **while** $\mathcal{Q}_{in}$ or $\mathcal{Q}_{out}$ are non-empty **do**
6:   **if** $\mathcal{Q}_{in}$ has node with highest activation **then**
7:     Pop best v from $\mathcal{Q}_{in}$ and insert into X_{in}
8:     **if** $iscomplete(v)$ **then** $emit(v)$
9:     **if** $depth_v < d_{max}$ **then**
10:       $\forall u \in incoming[v] : exploreEdge(u, v)$, **if** $u \notin X_{in}$ insert it into $\mathcal{Q}_{in}$ with depth $depth_v + 1$
11:     **if** $v \notin X_{out}$ insert it into $\mathcal{Q}_{out}$
12:   **else if** $\mathcal{Q}_{out}$ has node with highest activation **then**
13:     Pop best u from $\mathcal{Q}_{out}$ and insert into X_{out}
14:     **if** $iscomplete(u)$ **then** $emit(u)$
15:     **if** $depth_u < d_{max}$ **then**
16:       $\forall v \in outgoing[u] : exploreEdge(u, v)$, **if** $v \notin X_{out}$ insert it into $\mathcal{Q}_{out}$ with depth $depth_u + 1$

---

to reach a node in S_i in the best known path. $dist_{u,i}$ stores the length of the best known path from u to a node in S_i. $depth_u$ stores the number of edges of node u from the nearest keyword node. A depth cutoff value d_{max} is used to prevent generation of answers that would be unintuitive due to excessive path lengths, and to ensure termination.

At each iteration of the algorithm (line 5), between the incoming and outgoing iterators, the one having the node with highest priority is scheduled for exploration. Exploring a node v in $\mathcal{Q}_{in}$ (resp. $\mathcal{Q}_{out}$) is done as follows: incoming (resp. outgoing) edges are traversed to propagate keyword-distance information and activation information from v to adjacent nodes,[1] and the node is moved from $\mathcal{Q}_{in}$ to X_{out} (resp. $\mathcal{Q}_{out}$ to X_{out}). Additionally, if the node is found to have been reached from all keywords, it is "emitted" to an output queue. Answers are output from the output queue when the algorithm decides that no answers with lower cost will be generated in future.

---

[1]The activation of a node v is defined for every keyword k_i. Let $a_{v,i}$ be the activation of v with respect to keyword k_i. The activation of v is then defined as the sum of its activations from each keyword, i.e., $a_v = \sum_{i=1}^{l} a_{v,i} . a_{v,i}$ is first initialized for each node v that contains keyword k_i, and will spread to other nodes u to reflect the path length from u to keyword node k_i and how close u is to the potential root.

3.4.2   BI-LEVEL INDEXING

He et al. [2007] propose a bi-level index to speed up BIDIRECTIONALSEARCH, as no index (except the keyword-node index) is used in the original algorithm. A naive index precomputes and indexes all the distances from the nodes to keywords, but this will incur very large index size, as the number of distinct keywords is in the order of the size of the data graph G_D. A bi-level index can be built by first partitioning graph, and then building intra-block index and block index. Two node-based partitioning methods are proposed to partition a graph into blocks, namely, BFS-Based Partitioning, and METIS-Based Partitioning. Before introducing the data structure of the index, we first introduce the concept of a *portal*. In a node-based partitioning of a graph, a node separator is called a *portal* node (or portal for short). A block consists of all nodes in a partition as well as all portals incident to the partition. For a block, a portal can be either "in-portal", "out-portal", or both. A portal is called in-portal if it has at least one incoming edge from another block and at least one outgoing edge in this block. And a portal is called out-portal if it has at least one outgoing edge to another block and at least one incoming edge from this block.

For each block b, the intra-block index (IB-index) consists of the following data structures:

- **Intra-block keyword-node lists:** For each keyword k, $L_{KN}(b, k)$ denotes the list of nodes in block b that can reach k without leaving b, sorted according to their shortest distances (within b) to k (or more precisely, any node in b containing k).

- **Intra-block node-keyword map:** Looking up a node $u \in b$ together with a keyword k in this hash map returns $M_{NK}(b, u, k)$, the shortest distance (within b) from u to k (∞ if u cannot reach k in b).

- **Intra-block portal-node lists:** For each out-portal p of b, $L_{PN}(b, p)$ denotes the list of nodes in b that can reach p without leaving b, sorted according to shortest distances (within b) to p.

- **Intra-block node-portal distance map:** Looking up a node $u \in b$ in this hash map returns $D_{NP}(b, u)$, the shortest distance (within b) from a node u to the closest out-portal of b (∞ if u cannot reach any out-portal of b).

The primary purpose of L_{PN} is to support cross-block backward expansion in an efficient manner, as an answer may span multiple blocks through portals. The D_{NP} map gives the shortest distance between a node and its closest out-portal within a block. The block index is a simple data structure consisting of:

- **keyword-block lists:** For each keyword k, $L_{KB}(k)$ denotes the list of blocks containing keyword k, i.e., at least one node in b contains k;

- **portal-block lists:** For each portal p, $L_{PB}(p)$ denotes the list of blocks with p as an out-portal.

BLINKS [He et al., 2007], shown in Algorithm 26, works as follows. A priority queue Q_i of cursors is created for each keyword term k_i to simulate Dijkstra's algorithm by utilizing the distance

---

Algorithm 26 *BLINKS* (G_D, Q)

Input: a data graph G_D, and an l-keyword query $Q = \{k_1, k_2, \cdots, k_l\}$.
Output: directed rooted trees in increasing weight order.

1: **for** each $i \in [1, l]$ **do**
2:    $Q_i \leftarrow$ new Queue(); $\forall b \in L_{KB}(k_i) : Q_i.$INSERT(new Cursor($L_{KN}(b, k_i), 0$))
3: **while** $\exists j \in [1, l] : Q_j \neq \emptyset$ **do**
4:    $i \leftarrow$ pickKeyword($Q_1, \cdots, Q_l$)
5:    $c \leftarrow Q_i.$POP(); $\langle u, d \rangle \leftarrow c.$next()
6:    visitNode(i, u, d)
7:    **if** $\neg crossed(i, u)$ and $L_{PB}(u) \neq \emptyset$ **then**
8:       $\forall b \in L_{PB}(u) : Q_i.$INSERT(new Cursor($L_{PN}(b, u), d$))
9:       $crossed(i, u) \leftarrow true$
10:    $Q_i.$INSERT(c), if $c.$peekDist() $\neq \infty$
11:    **if** $|A| \geq K$ and $\sum_j Q_j.$TOP()$.peekDist() < \tau_{prune}$ and $\forall v \in R - A : sumLBDist(v) > \tau_{prune}$ **then**
12:       exit and output the top K answers in A
13: output up to top K answers in A

14: **Procedure** visitNode(i, u, d)
15: **if** $R[u] \neq \bot$ **then**
16:    $R[u] \leftarrow \langle u, \bot, \cdots, \bot \rangle; R[u].dist_i \leftarrow d$
17:    $b \leftarrow$ the block containing u
18:    **for** each $j \in [1, i) \cup (i, l]$ **do**
19:       $R[u].dist_i \leftarrow M_{NK}(b, u, k_i)$, if $D_{NP}(b, u) \geq M_{NK}(b, u, w_i)$
20: **else if** $sumLBDist(u) > \tau_{prune}$ **then**
21:    **return**
22: $R[u].dist_i \leftarrow d$
23: **if** $sumDist(u) < \infty$ **then**
24:    $A.$add($R[u]$)
25:    $\tau_{prune} \leftarrow$ the k-th largest of $\{sumDist(v)|v \in A\}$

---

information stored in the IB-index. Initially, for each keyword k_i, all the blocks that contain it are found by the keyword-block list, and a cursor is created to scan each intra-block keyword-node list and put in queue Q_i (line 3). The main part of the algorithm performs backward search, and it only conducts forward check at line 23. When an in-portal u is visited, all the blocks that have u as their out-portal need to be expanded (line 10) because a shorter path may cross several blocks.

The $pickKeyword(Q_1, \cdots, Q_l)$ chooses the next keyword (queue) to expand. When a keyword k_i visits node u for the first time (Procedure visitNode()), the distance d is guaranteed to be the shortest distance between u and k_i. The intra-block index $D_{NP}(b, u)$ and $M_{NK}(b, u, k_i)$ can be used to lower bound the shorted distance between u and k_i, if $D_{NP}(b, u) \geq M_{NK}(b, u, k_i)$, then the shortest distance is guaranteed to be $M_{NK}(b, u, k_i)$ (line 23); otherwise, it is lower bounded by $D_{NP}(b, u)$. One important optimization is that each keyword can across a portal at most one time (lines 9,11), i.e., any path that crosses the same portal more than one time can not be a shortest path. This reduces the search space dramatically.

Another index, called structure-aware index, is proposed to find answer for keyword query efficiently [Li et al., 2009b]. A different semantics of answer is defined, called *compact steiner tree*. The compact steiner tree is similar to the distinct root-based semantics. There is at most one answer tree, rooted at each node, for a keyword query. The tree, rooted at node t, is chosen as follows: for each keyword k_i, a node v containing k_i and dominating t (i.e., v is the node with shortest distance among all the nodes containing k_i that t can reach), is chosen, and the shortest path from t to v is added to the tree. Based on such a definition, the path from t to k_i becomes query independent, and it can be therefore precomputed and stored on disk. When a keyword query comes, it selects all the paths and joins them to form compact steiner trees.

3.4.3 EXTERNAL MEMORY DATA GRAPH

Dalvi et al. [2008] study keyword search on graphs where the graph G_D can not fit into main memory. They build a much smaller supernode graph on top of G_D that can resident in main memory. The supernode graph is defined as follows:

- **SuperNode:** The graph G_D is partitioned into components by a clustering algorithm, and each cluster is represented by a node called the *supernode* in the top-level graph. Each supernode thus contains a subset of $V(G_D)$, and the contained nodes (nodes in G_D) are called *innernodes*.

- **SuperEdge:** The edges between the supernodes called superedges are constructed as follows: if there is at least one edge from an innernode of supernode s_1 to an innernode of supernode s_2, then there exists a superedge from s_1 to s_2.

During supernode graph construction, the parameters are chosen such that the supernode graph fits into the available amount of main memory. Each supernode has a fixed number of innernodes and is stored on disk.

A *multi-granular graph* is used to exploit information presented in lower-level nodes (innernodes) that are cache-resident at the time a query is executed. A multi-granular graph is a hybrid graph that contains both supernodes and innernodes. A supernode is present either in *expanded* form, i.e., all its innernodes along with their adjacency lists are present in the cache, or in *unexpanded* form, i.e., its innernodes are not in the cache. The innernodes and their adjacency lists are handled in the unit of supernodes, i.e., either all or none of the innernodes of a supernode are presented in the cache. Since supernodes and innernodes coexist in the multi-granular graph, several types of edges can be present. Among these, the edges between supernodes and between innernodes need to be stored; the other edges can be inferred, i.e., the edges between supernodes are stored in main memory, and the edges between innernodes are stored in cache or on disk in the form of adjacency lists; the edges between supernode and innernode do not need to be stored explicitly. The weight of different kinds of edges are defined as follows.

- **supernode → supernode (S → S):** The edge weight of $s_1 \rightarrow s_2$ is defined as the minimum weight of those edges between the innernodes of s_1 and that of s_2, i.e., $w_e((s_1, s_2)) = \min_{v_1 \in s_1, v_2 \in s_2} w_e((v_1, v_2))$, where weight of edge (v_1, v_2) is defined to be ∞ if it does not exist.

- **supernode → innernode (S → I):** The edge weight of $s_1 \rightarrow v_2$ is defined as $w_e((s_1, v_2)) = \min_{v_1 \in s_1} w_e((v_1, v_2))$. These edges need not necessarily be explicitly represented. During the graph traversal, if s_1 is an unexpanded supernode, and there is a supernode s_2 in the adjacency list of supernode s_1, and s_2 is expanded, such edges can be enumerated by locating all innernodes $\{v_2 \in s_2|$ the adjacency list of v_2 contains some inner node in $s_1\}$.

- **innernode → supernode (I → S):** The edge weight in this case is defined in an analogous fashion to the previous case.

- **innernode → innernode (I → I):** Edge weight is the same as in the original graph.

When searching the multi-granular graph, the answers generated may contain supernodes, called *supernode answer*. If an answer does not contain any supernodes, it is called *pure answer*. The final answer returned to users must be pure answer. The Iterative Expansion Search algorithm (IES) [Dalvi et al., 2008] is a multi-stage algorithm that is applicable to mulit-granular graphs, as shown in Algorithm 27. Each iteration of IES can be broken up into two phases.

- **Explore phase:** Run an in-memory search algorithm on the current state of the multi-granular graph. The multi-granular graph is entirely in memory, whereas the supernode graph is stored in main memory, and details of expanded supernodes are stored in cache. When the search reaches an expanded supernode, it searches on the corresponding innernodes in cache.

- **Expand phase:** Expand the supernodes found in top-n ($n > k$) results of the previous phase and add them to input graph to produce an expanded multi-granular graph, by loading all the corresponding innernodes into cache.

The graph produced at the end of Expand phase of iteration i acts as the graph for iteration $i + 1$. Any in-memory graph search algorithm can be used in the Explore phase that treats all nodes (unexpanded supernode and innernode) in the same way. The multi-granular graph is maintained as a "virtual memory view", i.e., when visiting an expanded supernode, the algorithm will lookup its expansion in the cache, and load it into the cache if it is not in the cache. The algorithm stops when all top-k results are pure. Other termination heuristics can be used to reduce the time taken for query execution, at the potential cost of missed results.

Algorithm 27 restarts search (explore phase) every time from the scratch, which can lead to significantly increased CPU time. Dalvi et al. [2008] propose an alternative approach, called

---

Algorithm 27 Iterative Expansion Search(G, Q)

Input: a multi-granular graph G, and an l-keyword query $Q = \{k_1, k_2, \cdots, k_l\}$.
Output: top-k pure results.

1: **while** stopping criteria not satisfied **do**
2:     /* Explore phase */
3:     Run any in-memory search algorithm on G to generate the top-n results
4:     /* Expand phase */
5:     **for** each result R in top-n results **do**
6:         $SNodeSet \leftarrow SNodeSet \cup \{$all super nodes from $R\}$
7:     Expand all supernodes in $SNodeSet$ and add them to G
8: output top-k pure results

Algorithm 28 Iterative Expansion Backward Search(G, Q)

Input: a multi-granular graph G, and an l-keyword query $Q = \{k_1, k_2, \cdots, k_l\}$.
Output: top-k pure results.

1: **while** less than k pure results generated **do**
2:     $Result \leftarrow$ BackwardSearch.GetNextResult()
3:     **if** $Result$ contains a supernode **then**
4:         Expand one or more supernodes in $Result$ and update the SPI trees that contain those expanded supernodes
5: output top-k pure results

---

incremental expansion. When a supernode answer is generated, one or more supernodes in the answer are expanded. However, instead of restarting each time when supernodes are expanded, incremental expansion updates the state of the search algorithm. Once the state is updated, search continues from where it left off earlier, on the modified graph. Algorithm 28 shows the Incremental Expansion Backward search (IEB) where the in-memory search is implemented by a backward search algorithm. There is one shortest path iterator (SPI) tree per keyword k_i, which contains all nodes "touched" by Dijkstra's algorithm, including explored nodes and fringe nodes, starting from k_i (or more precisely S_i). More accurately, the SPI tree does not contain graph nodes, rather each tree-node of an SPI tree contains a pointer to a graph node. From the SPI tree, the shortest path from an explored node to an keyword node can be identified. The backward search algorithm expands each SPI tree using Dijkstra's algorithm. When an answer is output by the backward search algorithm, if it contains any supernode, one or more supernodes from the answer are expanded, otherwise it is output. When a supernode is expanded, the SPI trees that contain this supernode should be updated to include all the innernodes and exclude this supernode.

3.5   SUBGRAPH-BASED KEYWORD SEARCH

The previous sections define the answer of a keyword query as Q-SUBTREE, which is a directed subtree. We show two subgraph-based notions of answer definition for a keyword query in the following, namely, r-radius steiner graph, and multi-center induced graph.

3.5.1 r-RADIUS STEINER GRAPH

Li et al. [2008a] define the result of an l-keyword query as an r-radius steiner subgraph. The graph is unweighted and undirected, and the length of a path is defined as the number of edges in it. The definition of r-radius steiner graph is based on the following concepts.

Definition 3.11   Centric Distance.   Given a graph G and any node $v \in V(G)$, the centric distance of v in G, denoted as $CD(v)$, is the maximum among the shortest distances between v and any node $u \in V(G)$, i.e., $CD(v) = \max_{u \in V(G)} dist(u, v)$.

Definition 3.12   Radius.   The radius of a graph G, denoted as $\mathcal{R}(G)$, is the minimum value among the centric distances of every node in G, i.e., $\mathcal{R}(G) = \min_{v \in V(G)} CD(v)$. G is called an r-radius graph if its radius is exactly r.

Definition 3.13 r-Radius Steiner Graph.   Given an r-radius graph G and a keyword query Q, node v in G is called a content node if it contains some of the input keywords. Node s is called steiner node if there exist two content nodes, u and v, and s in on the simple path between u and v. The subgraph of G composed of the steiner nodes and associated edges is called an r-radius steiner graph (SG). The radius of an r-radius steiner graph can be smaller than r.

Example 3.14   Figure 3.9(a) shows two subgraphs, SG_1 and SG_2, of the data graph shown in Figure 3.1(e). In SG_1, the centric distance of t_1 and t_8 are $CD(t_1) = 2$ and $CD(t_8) = 3$, respectively. In SG_2, the centric distance of t_1 and t_8 are $CD'(t_1) = 3$ and $CD'(t_8) = 3$, respectively. The radius of SG_1 and SG_2 are $\mathcal{R}(SG_1) = 2$ and $\mathcal{R}(SG_2) = 3$, respectively. For a keyword query $Q = \{Brussels, EU\}$, one 2-radius steiner graph is shown in Figure 3.9(b), where t_6 contains keyword "Brussels" and t_3 contains keyword "EU", and it is obtained by removing the non-steiner nodes from SG_1.

Note that the definition of r-radius steiner graph is based on r-radius subgraph. A more general definition of r-radius steiner graph would be any induced subgraph satisfying the following two properties: (1) the radius should be no more than r, (2) every node should be either a content node or a steiner node. The actual problem of a keyword query in this setting is to find r-radius subgraphs, and the corresponding r-radius steiner graph is obtained as a post-processing step as described by the definition.

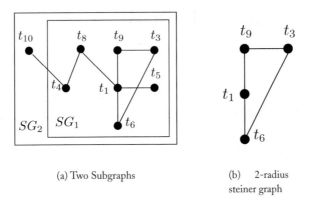

(a) Two Subgraphs

(b)   2-radius steiner graph

Figure 3.9: 2-radius steiner graph for $Q = \{\text{Brussels, EU}\}$

The approaches to find r-radius subgraphs are based on the adjacency matrix, $M = (m_{ij})_{n \times n}$, with respect to G_D, which is a $n \times n$ Boolean matrix. An element m_{ij} is 1, if and only if there is an edge between v_i and v_j, m_{ii} is 1 for all i. $M^r = M \times M \cdots \times M = (m_{ij})_{n \times n}$ is the r-th power of adjacency matrix M. An element m_{ij}^r is 1, if and only if the shortest path between v_i and v_j is less than or equal to r. $N_i^r = \{v_j | m_{ij}^r = 1\}$ is the set of nodes that have a path to v_i with distance no larger than r. G_i^r denotes the subgraph induced by the node set N_i^r. $G_{v_i}^r (N_{v_i}^r)$ can be interchangeably used instead of $G_i^r(N_i^r)$. We use $G_i \trianglelefteq G_j$ to denote that G_i is a subgraph of G_j. The r-radius subgraph is defined based on G_i^r's. The following lemma is used to find all the r-radius subgraphs [Li et al., 2008a].

Lemma 3.15   *[Li et al., 2008a] Given a graph G, with $\mathcal{R}(G) \geq r > 1$, $\forall i$, $1 \leq i \leq |V(G)|$, G_i^r is an r-radius subgraph, if, $\forall v_k \in N_i^r$, $N_i^r \nsubseteq N_k^{r-1}$.*

Note that, the above lemma is a sufficient condition for identifying r-radius subgraphs, but it is not a necessary condition. In principle, there can be, exponentially, many r-radius subgraphs of G. Li et al. [2008a] only consider $n = |V(G)|$ subgraphs; each is uniquely determined by one node in G, while other r-radius subgraphs are possible.

An r-radius subgraph G_i^r is maximal if and only if there is no other r-radius subgraph G_j^r that is a super graph of G_i^r, i.e. $G_i^r \trianglelefteq G_j^r$. Li et al. [2008a] consider those maximal r-radius subgraphs G_i^r as the subgraphs that will generate r-radius steiner subgraphs. All these maximal r-radius subgraphs G_i^r are found, which can be pre-computed and indexed on the disk, because these maximal r-radius graph are query independent.

The objective here is to find top-k r-radius steiner subgraphs, and ranking functions are introduced to rank the r-radius steiner subgraphs. Each keyword term k_i has an IR-style score:

$$Score_{IR}(k_i, SG) = \frac{ntf(k_i, G) \times idf(k_i)}{ndl(G)}$$

that is a normal TF-IDF score, where $idf(k_i)$ indicates the relative importance of keyword k_i and $ntf(k_i, G)$ measures the relevance of G to keyword k_i. Here G is the subgraph from which the r-radius steiner subgraph is generated. Each keyword pair (k_i, k_j) has a structural score, which measures the compactness of the two keywords in SG.

$$Sim(\langle k_i, k_j \rangle | SG) = \frac{1}{|C_{k_i} \cup C_{k_j}|} \sum_{v_i \in C_{k_i}, v_j \in C_{k_j}} Sim(\langle v_i, v_j \rangle | SG)$$

where $C_{k_i}(C_{k_j})$ is the set of keyword nodes in SG that contain $k_i(k_j)$, and $Sim(\langle v_i, v_j \rangle | SG) = \sum_{p \in path(\langle v_i, v_j \rangle | SG)} \frac{1}{(len(p)+1)^2}$, where $path(\langle v_i, v_j \rangle | SG)$ denote the set of all the paths between v_i and v_j in SG and $len(p)$ is the length of path p. Intuitively, $Sim(\langle k_i, k_j \rangle | SG)$ measures how close the two keywords, k_i and k_j, are connected to each other. The final score of SG is defined as follows,

$$Score(\{k_1, \cdots, k_l\}, SG) = \sum_{1 \leq i < j \leq l} Score(\langle k_i, k_j \rangle | SG)$$

where

$$Score(\langle k_i, k_j \rangle | SG) = Sim(\langle k_i, k_j \rangle | SG) \times (Score_{IR}(k_i, SG) + Score_{IR}(k_j, SG))$$

According to the definition of r-radius steiner subgraph, $Score(\langle k_i, k_j \rangle | SG)$ can be directly computed on G. Then $Score(\langle k_i, k_j \rangle | SG)$ can be pre-computed for each possible keyword pair and each maximal r-radius subgraph G_i^r. An index can be built by storing a list of maximal r-radius subgraphs G in decreasing order of $Score(\langle k_i, k_j \rangle | SG)$, for each possible keyword pair. When a keyword query arrives, it can directly use these lists, and by applying the *Threshold Algorithm* [Fagin, 1998] the top-k maximal r-radius subgraphs can be obtained, and then the top-k r-radius steiner subgraphs can be computed by refining the corresponding r-radius subgraphs.

3.5.2   MULTI-CENTER INDUCED GRAPH

In contrast to tree-based results that are single-center (root) induced trees, in this section, we consider query answers that are multi-centered induced subgraphs of G_D. These are referred to as *communities* [Qin et al., 2009b]. The vertices of a community $R(V, E)$, $V(R)$ is a union of three subsets, $V = V_c \cup V_l \cup V_p$, where V_l represents a set of keyword nodes (*knode*), V_c represents a set of center nodes (*cnode*) (for every *cnode* $v_c \in V_c$, there exists at least a single path such that $dist(v_c, v_l) \leq R_{max}$ for any $v_l \in V_l$, where R_{max} is introduced to control the size of a community), and V_p represents a

---

Algorithm 29 GetCommunity(G_D, C, R_{max})

Input: a data graph G_D, a core $C = [c_1, \cdots, c_l]$, and a radius threshold R_{max}.
Output: A community uniquely determined by C.

1: Find the set of *cnodes*, V_c, by running $|C|$ copies of Dijkstra's single source shortest path algorithm
2: Run a single copy of Dijkstra's algorithm to find the shortest distance to the nearest *knode*, for each node $v \in V(G_D)$, i.e. $dist_k(v) = \min_{c \in C} dist(v, c)$
3: Run a single copy of Dijkstra's algorithm to find the shortest distance from the nearest *cnode*, for each node $v \in V(G_D)$, i.e. $dist_c(v) = \min_{v_c \in V_c} dist(v_c, v)$
4: $\mathcal{V} \leftarrow \{u \in V(G_D) | dist_c(u) + dist_k(u) \leq R_{max}\}$
5: Construct a subgraph $\mathcal{R}$ in G_D induced by $\mathcal{V}$ and return it

---

set path nodes (*pnode*) that include all the nodes that appear on any path from a *cnode* $v_c \in V_c$ to a *knode* $v_l \in V_l$ with $dist(v_c, v_l) \leq R_{max}$. $E(R)$ is the set of edges induced by $V(R)$.

A community, R, is uniquely determined by the set of *knodes*, V_l, which is called the core of the community and denoted as $core(R)$. The weight of a community R, $w(R)$ is defined as the minimum value among the total edge weights from a *cnode* to every *knode*; more precisely,

$$w(R) = \min_{v_c \in V_c} \sum_{v_l \in V_l} dist(v_c, v_l). \tag{3.8}$$

For simplicity, we use C to represent a core as a list of l nodes, $C = [c_1, c_2, \cdots, c_l]$, and it may use $C[i]$ to denote $c_i \in C$, where c_i contains the keyword term k_i. Based on the definition of community, once the core C is provided, the community is uniquely determined, and it can be found by Algorithm 29, which is self-explanatory.

Qin et al. [2009b] enumerate all (or the top-k) communities in polynomial delay by adopting the Lawler's procedure [Lawler, 1972]. The general idea is the same as ENUMTREEPD (Algorithm 19). But it is much easier here, because ENUMTREEPD enumerates trees which has structure, while in this case only the cores are enumerated where each core is just a set of l keyword nodes. In this problem, the answer space is $S_1 \times S_2 \cdots \times S_l$, where each S_i is the set of nodes in G_D that contains keyword k_i. A subspace is described by $V_1 \times V_2 \cdots, \times V_l$ where $V_i \subseteq S_i$ and it also can be compactly described by a set of inclusion constraints and exclusion constraints. Based on Lawler's procedure, in order to enumerate the communities in increasing cost order, it is straightforward to obtain an algorithm whose time complexity of delay is $O(l \cdot c(l))$, where $c(l)$ is the time complexity to compute the best community.

Two algorithms are proposed for enumerating communities in order with time complexity $O(c(l))$: one enumerates all communities in arbitrary order with polynomial delay, and the other enumerates top-k communities in increasing weight order with polynomial delay. In the following, we discuss the second algorithm.

Algorithm 30 COMM-K(G_D, Q, R_{max})

Input: a data graph G_D, keywords set $Q = \{k_1, \cdots, k_l\}$, and a radius threshold R_{max}.
Output: Enumerate top-K communities in increasing weight order.

1: Find the set of *knodes* $\{S_1, \cdots, S_l\}$ and their corresponding neighborhood nodes $\{N_1, \cdots, N_l\}$
2: Find the best core (with lowest weight) and the corresponding weight from $\{N_1, \cdots, N_l\}$, denoted $(C, weight)$
3: Initialize $\mathcal{H} \leftarrow \emptyset$; $\mathcal{H}$.INSERT($C, weight, 1, \emptyset$)
4: **while** $\mathcal{H} \neq \emptyset$ **and** less than K communities output **do**
5:     $g \leftarrow \mathcal{H}$.POP(); $\{g = (C, weight, pos, prev)\}$
6:     $R' \leftarrow GetCommunity(G_D, g.C, R_{max})$, and output R'
7:     $\forall i \in [1, l]$: update N_i to be the neighborhood nodes of $g.C[i]$, $V_i \leftarrow S_i$
8:     update $\{V_1, \cdots, V_l\}$ by following the links $g.prev$ recursively
9:     **for** $i = l$ **downto** $g.pos$ **do**
10:         $V_i \leftarrow V_i - \{g.C[i]\}$, update N_i to be the neighborhood nodes of V_i
11:         Find the best core from the current $\{N_1, \cdots, N_l\}$, denoted $(C', weight')$
12:         $\mathcal{H}$.INSERT($C', weight', i, g$) **if** C' exists
13:         $V_i \leftarrow V_i \cup \{g.C[i]\}$, update N_i to be the neighborhood nodes of V_i

---

Algorithm 30 shows the high-level pseudocode. $\mathcal{H}$ is a priority heap, used to store the intermediate and potential cores with additional information. The general idea is to consider the entire set of potential cores as an l-dimensional space $S_1 \times S_2 \cdots \times S_l$, and at each step, divide a subspace into smaller subspaces and find a best core in each newly generated subspace. At any intermediate step, the whole set of subspaces are disjoint, and the union is guaranteed to cover the whole space. Each time a core with the lowest weight is removed from $\mathcal{H}$, it is guaranteed to be the next community in order (line 5). The best core of a subspace $V_1 \times V_2 \cdots \times V_l$, where $V_i \subset S_i$, is found as follows (lines 2,11). First, a neighborhood nodeset N_i is found for each set V_i, which consists of all the nodes with a shortest distance no greater than R_{max} to at least one of the nodes in V_i. This can be done by running a shortest path algorithm. Second, a linear scan of the nodes can find the best core with the best center and weight. When the next best core $g.C$ is found, the subspace from which $g.C$ is found is partitioned into several subspaces (lines 9-13); the best core from each newly generated subspace is found (line 11) and inserted into $\mathcal{H}$ (line 12). Each entry in $\mathcal{H}$ consists of four fields, $(C, weight, pos, prev)$, where C is the core and $weight$ is the corresponding weight, pos and pre is used to reconstruct efficiently the subspace (without storing the description of the subspace explicitly) from which C is computed.

Algorithm 30 enumerates top-k communities in increasing weight order, with time complexity $O(l(n \log n + m))$, and using space $O(l^2 \cdot k + l \cdot n + m)$ [Qin et al., 2009b]. Note that, finding the best core in a subspace (under inclusion constraints and exclusion constraints) also takes time $c(l) = O(l(n \log n + m))$. According to discussion of ENUMTREEPD, it is easy to get an enumeration algorithm with delay $l \cdot c(l)$. However, information can be shared during consecutive execution of Line 11 of EnumTreePD, so Algorithm 30 can enumerate communities with delay $c(l)$.

CHAPTER 4

Keyword Search in XML Databases

In this chapter, we focus on keyword search in *XML* databases where an *XML* database is treated as a large data tree. We introduce various semantics to answer a keyword query on *XML* tree, and we discuss efficient algorithms to find the answers under such semantics. A main difference between this chapter and the previous chapters is that the underlying data structure is a large tree instead of a large graph.

In Section 4.1, we introduce several important concepts and definitions such as Lower Common Ancestor (LCA), Smallest LCA (SLCA), Exclusive LCA (ELCA), and Compact LCA (CLCA). Their properties and the relationships among LCA, SLCA and ELCA will be discussed. In Section 4.2, we discuss the algorithms that find answers based on SLCA. In Section 4.3, we discuss the algorithms that focus on identifying meaningful return information. We discuss algorithm to find answers based on ELCA in Section 4.4. In Section 4.5, in brief, we give several approaches based on meaning LCA, *interconnection*, and relevance oriented ranking.

4.1 XML AND PROBLEM DEFINITION

XML is modeled as a rooted and labeled tree, such as the one shown in Figure 4.1. Each internal node v in the tree corresponds to an *XML* element, called *element node*, and is labeled with a tag/label $tag(v)$. Each leaf node of the tree corresponds to a data value, called *value node*. For example, in Figure 4.1, "Dean" and "Title" are element nodes, "John" and "Ben" are value nodes. In this model, the attribute nodes are modeled as children of the associated element node, and we do not distinguish them from element nodes.

Each node (element node or value node) in the *XML* tree is assigned an unique *Dewey ID*. The *Dewey ID* of nodes are assigned in the following way: the relative position of each node among its siblings are recorded, and the concatenation of these relative positions using dot '.' starting from the root composes the *Dewey ID* of the nodes. For example, the node with *Dewey ID* 0.1.2.1 (Students) is the second child of its parent node 0.1.2 (Class). We denote the *Dewey ID* of a node v as $pre(v)$, as it is compatible with the preorder numbering, i.e., a node v_1 precedes another node v_2 in the preorder left-to-right depth-first traversal of the tree, if and only if $pre(v_1) < pre(v_2)$. The $<$ relationship between two *Dewey IDs* is the same as comparing between two sequences. Besides the order information preserved by the *Dewey ID*, it also can be used to detect sibling and ancestor-descendant relationships between nodes.

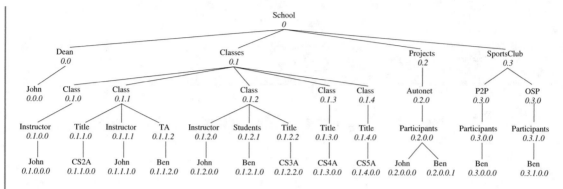

Figure 4.1: Example *XML* documents [Xu and Papakonstantinou, 2005]

- A node *u* is a sibling of node *v* if and only if *pre(u)* differs from *pre(v)* only in the last component. For example, 0.1.1.0 (Title) and 0.1.1.1 (Instructor) are sibling nodes, but 0.1.1 (Class) and 0.1.1.1 (Instructor) are not sibling nodes.

- A node *u* is an ancestor of another node *v* if and only if *pre(u)* is a prefix of *pre(v)*. For example, 0.1 (Classes) is an ancestor of 0.1.2.0.0 (John).

For simplicity, we use $u < v$ to denote that $pre(u) < pre(v)$. $u \leq v$ denotes that $u < v$ or $u = v$. We also use $u \prec v$ to denote that u is an ancestor of v, or equivalently, v is a descendant of u. $u \preceq v$ denotes that $u \prec v$ or $u = v$. Note that, if $u \prec v$ then $u < v$, but the other direction is not always true.

4.1.1   LCA, SLCA, ELCA, AND CLCA

In the following, we show the definitions of LCA, SLCA [Xu and Papakonstantinou, 2005], ELCA [Guo et al., 2003], and CLCA [Li et al., 2007a], which are the basis of semantics of answer definitions.

Definition 4.1   Lowest Common Ancestor (LCA).   For any two nodes v_1 and v_2, u is the LCA of v_1 and v_2 if and only if: (1) $u \prec v_1$ and $u \prec v_2$, (2) for any u', if $u' \prec v_1$ and $u' \prec v_2$, then $u' \preceq u$. The LCA of nodes v_1 and v_2 is denoted as $lca(v_1, v_2)$. Note that $lca(v_1, v_2)$ is the same as $lca(v_2, v_1)$.

Property 4.2   Given any three nodes v_2, v_1, v where $v_2 < v_1 < v$, $lca(v, v_2) \preceq lca(v, v_1)$. Given any three nodes v, v_1, v_2 where $v < v_1 < v_2$, $lca(v, v_2) \preceq lca(v, v_1)$.

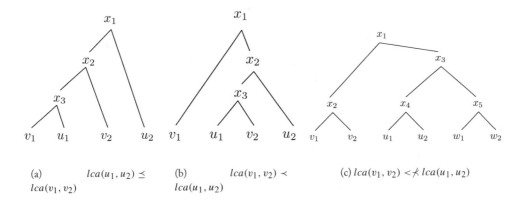

Figure 4.2: Different situations of $lca(v_1, v_2)$ and $lca(u_1, u_2)$

Property 4.3   Given any two pairs of nodes (v_1, v_2) and (u_1, u_2), with $v_1 \leq u_1$ and $v_2 \leq u_2$, without loss of generality, we can assume that $v_1 < v_2$ and $u_1 < u_2$. Let $lca(v_1, v_2)$ and $lca(u_1, u_2)$ be the LCA of v_1, v_2 and u_1, u_2, respectively. Then,

1. if $lca(v_1, v_2) \geq lca(u_1, u_2)$, then $lca(u_1, u_2) \preceq lca(v_1, v_2)$, as shown in Figure 4.2(a),

2. if $lca(v_1, v_2) < lca(u_1, u_2)$, then

    - either $lca(v_1, v_2) \prec lca(u_1, u_2)$, as shown in Figure 4.2(b),
    - or $lca(v_1, v_2) \nprec lca(u_1, u_2)$, in which case for any w_1, w_2 with $u_1 \leq w_1$ and $u_2 \leq w_2$, $lca(v_1, v_2) \nprec lca(w_1, w_2)$, as shown in Figure 4.2(c).

The above definition of LCA for two nodes can be straightforwardly extended to the definition of LCA for more than two nodes. Let $lca(v_1, \cdots, v_l)$ denote the LCA of nodes $v_1, \cdots, v_l$, where $lca(v_1, \cdots, v_l) = lca(lca(v_1, \cdots, v_{l-1}), v_l)$ for $l > 2$. The LCA of sets of nodes, $S_1, \cdots, S_l$, is the set of LCA for each combination of nodes in S_1 through S_l; more precisely,

$$lca(S_1, \cdots, S_l) = \{lca(v_1, \cdots, v_l) \mid v_1 \in S_1, \cdots, v_l \in S_l\}$$

For example, in Figure 4.1, let S_1 be the set of nodes containing keyword "John", i.e., $S_1 = \{0.0.0, 0.1.0.0.0, 0.1.1.1.0, 0.1.2.0.0, 0.2.0.0.0\}$, and let S_2 be the set of nodes containing keyword "Ben,", i.e., $S_2 = \{0.1.1.2.0, 0.1.2.1.0, 0.2.0.0.1, 0.3.0.0.0, 0.3.1.0.0\}$. Then $lca(S_1, S_2) = \{0, 0.1, 0.1.1, 0.1.2, 0.2, 0.2.0, 0.2.0.0\}$

Definition 4.4 Smallest LCA (SLCA).   The SLCA of l sets $S_1, \cdots, S_l$ is defined to be

$$slca(S_1, \cdots, S_l) = \{v \in lca(S_1, \cdots, S_l) \mid \forall v' \in lca(S_1, \cdots, S_l), v \nprec v'\}.$$

Intuitively, it is the set of nodes in $lca(S_1, \cdots, S_l)$ such that none of their descendants is in $lca(S_1, \cdots, S_l)$.

A node v is called a SLCA of $S_1, \cdots, S_l$ if $v \in slca(S_1, \cdots, S_l)$. Note that a node in $slca(S_1, \cdots, S_l)$ can not be an ancestor of any other node in $slca(S_1, \cdots, S_l)$. Continuing the above example, where S_1 and S_2 are the set of nodes that contain keywords "John" and "Ben" respectively, the SLCA of S_1 and S_2 is $slca(S_1, S_2) = \{0.1.1, 0.1.2, 0.2.0.0\}$, namely, 0.1.1 (Class), 0.1.2 (Class), and 0.2.0.0 (Participants).

Definition 4.5   Exclusive LCA (ELCA).   The ELCA of l sets $S_1, \cdots, S_l$ is defined to be

$$elca(S_1, \cdots, S_l) = \quad \{u \mid \exists v_1 \in S_1, \cdots, v_l \in S_l, (u = lca(v_1, \cdots, v_l) \wedge$$
$$\forall i \in [1, l], \nexists x (x \in lca(S_1, \cdots, S_l) \wedge child(u, v_i) \preceq x))\}$$

where $child(u, v_i)$ is the child of u in the path from u to v_i.

A node u is called an ELCA of l sets $S_1, \cdots, S_l$ if $u \in elca(S_1, \cdots, S_l)$, i.e., if and only if there exist l nodes $v_1 \in S_1, \cdots, v_l \in S_l$, such that $u = lca(v_1, \cdots, v_l)$, and for every v_i $(1 \leq i \leq l)$ the child of u in the path from u to v_i is not in $lca(S_1, \cdots, S_l)$ nor ancestor of any node in $lca(S_1, \cdots, S_l)$. The node v_i is called an ELCA *witness node* of u in S_i. Note that, the witness node v_i of an ELCA node u can not be ancestor of any ELCA node. Intuitively, each ELCA node has a set of l witness nodes that are not descendants or ancestors of other ELCA nodes. For example, $elca(S_1, S_2) = \{0, 0.1.1, 0.1.2, 0.2.0.0\}$, the node 0.0.0 is an ELCA witness node of the node 0 in S_1, and the node 0.3.0.0.0 is an ELCA witness node of the node 0 in S_2.

Definition 4.6   Compact LCA (CLCA).   Given l nodes, $v_1 \in S_1, \cdots, v_l \in S_l$, $u = lca(v_1, \cdots, v_l)$. u is said to dominate v_i if $u = slca(S_1, \cdots, S_{i-1}, \{v_i\}, \cdots, S_l)$. u is a CLCA with respect to these l nodes, if and only if u dominates each v_i.

Given l set of nodes $S_1, \cdots, S_l$, let SLCAs and ELCAs denote the set of SLCA nodes $slca(S_1, \cdots, S_l)$ and ELCA nodes $elca(S_1, \cdots, S_l)$, respectively. Actually, the set of CLCA nodes is the same as ELCAs, as proven in the following theorem.

Theorem 4.7   *Given l nodes, $v_1 \in S_1, \cdots, v_l \in S_l$, $u = lca(v_1, \cdots, v_l)$ is a CLCA with respect to $v_1, \cdots, v_l$, if and only if $u \in elca(S_1, \cdots, S_l)$ with $v_1, \cdots, v_l$ as witness nodes.*

Proof.   First, we prove $\Rightarrow$ by contradiction. Let u be a CLCA w.r.t $v_1, \cdots, v_l$. Assume that u is not an ELCA with $v_1, \cdots, v_l$ as witness nodes, then there must exist a $i \in [1, l]$ and a $x \in lca(S_1, \cdots, S_l)$, with $child(u, v_i) \preceq x$. Then $child(u, v_i) \preceq slca(S_1, \cdots, S_{i-1}, \{v_i\}, \cdots, S_l)$, which means that u does not dominate v_i, then a contradiction found. Second, we prove $\Leftarrow$ by contradiction.

Let u be an ELCA with witness nodes $v_1, \cdots, v_l$. Assume that u is not a CLCA with respect to $v_1, \cdots, v_l$; then there must exist a $i \in [1, l]$ where u does not dominate v_i, i.e., $u \prec slca(S_1, \cdots, S_{i-1}, \{v_i\}, S_{i+1}, \cdots, S_l)$. Then $child(u, v_i) \preceq slca(S_1, \cdots, \{v_i\}, \cdots, S_l)$, which is a contradiction.   □

Theorem 4.8   *[Xu and Papakonstantinou, 2008] The relationship between LCA nodes,* SLCA *nodes, and* ELCA *nodes, of l sets $S_1, \cdots, S_l$, is $slca(S_1, \cdots, S_l) \subseteq elca(S_1, \cdots, S_l) \subseteq lca(S_1, \cdots, S_l)$.*

For example, consider S_1 and S_2 as the set of nodes containing keyword "John" and "Ben," respectively, the node 0 (School) is an ELCA node but not a SLCA node, and the node 0.1 (Classes) is a LCA node but not an ELCA node.

4.1.2   PROBLEM DEFINITION AND NOTATIONS

Given a list of l keywords $Q = \{k_1, \cdots, k_l\}$, and an input *XML* tree T, the problem is to find a set of meaningful subtrees defined by (t, M), i.e., $\mathcal{R}(T, Q) = \{(t, M)\}$. For each subtree (t, M), t is the root node of the subtree, and M are match nodes; it should have at least one match node for each keyword (i.e., a node is called a *match node* if it contains one keyword), and $t = lca(v_1, \cdots, v_m)$ (assume that $M = \langle v_1, \cdots, v_m \rangle$).

Different semantics have been proposed to define the meaningful subtrees, e.g., SLCA based, ELCA based, MLCA based [Li et al., 2004, 2008b] and *interconnection* [Cohen et al., 2003]. In most of the works in the literature, there exists an inverted index of *Dewey IDs* for each keyword. Using the inverted index, for an l-keyword query, it is possible to get l lists $S_1, \cdots, S_l$. Each S_i ($1 \le i \le l$) contains the set of nodes containing the keyword k_i, and the nodes contained in S_i are match nodes. Let $|S_i|$ denote the number of nodes in S_i. Without loss of generality, we assume that S_1 has the smallest cardinality among $S_1, \cdots, S_l$. Let $|S|$ denote the maximum value among the cardinality of S_i's, i.e., $|S| = \max_{1 \le i \le l} |S_i|$. The algorithms work on the l lists $S_1, \cdots, S_l$. Below, we also use $slca(Q)$ and $elca(Q)$ to denote $slca(S_1, \cdots, S_l)$ and $elca(S_1, \cdots, S_l)$, respectively. Note that lists S_i are sorted in increasing *Dewey ID* order. We assume that $S_i \ne \emptyset$ for $1 \le i \le l$.

In the following, we use d to denote the height of the *XML* tree, i.e., d is the maximum length of all the *Dewey IDs* of the nodes in the *XML* tree. Given two nodes u and v with their *Dewey IDs*, we can find $lca(u, v)$ in time $O(d)$, based on the fact that $lca(u, v)$ has a *Dewey ID* that is equal to the longest common prefix of $pre(u)$ and $pre(v)$. Note that $lca(u, v)$ exists for any two nodes in a tree, because both u and v are descendants of the root node. We define $lca(u, \bot)$ to be $\bot$, where $\bot$ denotes a *null* node (value). Note that the preorder and postorder relationships between u and $\bot$ are not defined.

We first discuss some primitive functions used by the algorithms that we will present later. Assume that each set S is sorted in increasing order of *Dewey ID*.

- $lm(v, S)$: computes the *left match* of v in a set S, which is the node in S that has the largest *Dewey ID* that is less than or equal to $pre(v)$, i.e. $lm(v, S) = \arg\max_{u \in S:u \le v} pre(u)$. It returns

$\perp$, when there is no left match node. The cost of the function is $O(d \log |S|)$, and it can be implemented by a binary search on S, which takes $O(\log |S|)$ steps, and each step takes $O(d)$ time to compare two *Dewey IDs*.

- $rm(v, S)$: computes the *right match* of v in a set S, which is the node in S that has the smallest *Dewey ID* that is greater than or equal to $pre(v)$, i.e. $rm(v, S) = \arg\min_{u \in S: u \geq v} pre(u)$. It returns $\perp$ when there is no right match node. The cost of the function is $O(d \log |S|)$.

- $closest(v, S)$: computes the closest node of v in S, which is either $lm(v, S)$ or $rm(v, S)$. When either one is $\perp$, then $closest(v, S)$ is defined as the other one; otherwise, $closest(v, S)$ is defined to be $lm(v, S)$, if $lca(v, rm(v, S)) \preceq lca(v, lm(v, S))$, $closest(v, S) = rm(v, S)$ otherwise. The cost of $closest(v, S)$ is $O(d \log |S|)$, where $lca()$ takes $O(d)$ time, $lm()$ and $rm()$ take $O(d \log |S|)$ time.

- $removeAncestor(S)$: returns the subset of nodes in S whose descendants are not in S, i.e. $removeAncestor(S) = \{v \in S \mid \nexists u \in S : v \prec u\}$. The cost of $removeAncestor$ is $O(d|S|)$, since S is sorted in increasing *Dewey ID* order.

With the *Dewey IDs*, comparing two nodes takes $O(d)$ time, and computing lca of two nodes also takes $O(d)$ time. Note that there exists another encoding for *XML* tree, called interval encoding, that stores three numbers for each node $\langle start, end, level \rangle$, where $start$ is the number assigned by a preorder traversal, end is the largest $start$ value among the nodes in the subtree rooted at that node, and $level$ is the level of the node in *XML* tree. Using interval encoding, comparing two nodes takes $O(1)$ time, i.e., it takes $O(1)$ time to determine the relationships of $u < v$, $u \prec v$ or u is the parent of v for two nodes u and v. But most of the works in the literature use only *Dewey ID* to encode nodes, so in the following, we only consider the *Dewey ID* encode, where comparing two nodes takes $O(d)$ time.

4.2　SLCA-BASED SEMANTICS

The intuition of SLCA-based semantics of keyword search is that, each node in T can be viewed as an entity in the world. If u is an ancestor of v, then we may understand that the entity represented by v belongs to the entity that u represents. For example, in Figure 4.1, the entity represented by 0.1.1 (Class) belongs to the entity represented by 0 (School). For a keyword query, it is more desirable to return the most specific entities that contain all the keywords, i.e., among all the returned entities, there should not exist any ancestor-descendant relationship between the root nodes t that represent entities.

In this section, we first show some properties of the *slca* function, which is essential for efficient algorithms. Then three efficient algorithms with different characteristics are shown to compute $slca(S_1, \cdots, S_l)$ for an l-keyword query. Even using SLCA-based semantics, different subtrees can be returned for a SLCA node; we will present several properties that the answers should have, based on which relevant subtrees for each SLCA node can be identified.

4.2.1   PROPERTIES OF LCA AND SLCA

Property 4.9   Given a set S and two nodes v_i and v_j with $v_i < v_j$, then $closest(v_i, S) \leq closest(v_j, S)$.

Proof. We prove it by contradiction, by assuming that $closest(v_i, S) > closest(v_j, S)$. Then $closest(v_i, S) = rm(v_i, S)$ and $closest(v_j, S) = lm(v_j, S)$, $rm(v_i, S) > lm(v_j, S)$. Recall that $closest(v, S)$ is chosen from $lm(v, S)$ and $rm(v, S)$, and $lm(v_i, S) \leq lm(v_j, S)$ and $rm(v_i, S) \leq rm(v_j, S)$ if all exists. If $lm(v_j, S) < rm(v_i, S)$, then $lm(v_j, S) \leq lm(v_i, S)$, therefore $lm(v_i, S) = lm(v_j, S)$ by the fact that $lm(v_i, S) \leq lm(v_j, S)$. Similarly, we can get that $rm(v_i, S) = rm(v_j, S)$. Also, we can learn that $lm(v_i, S) \neq rm(v_i, S)$, otherwise $closest(v_i, S) = lm(v_i, S)$.

Let lm denote $lm(v_i, S)$ and rm denote $rm(v_i, S)$. It holds that $lm < v_i < v_j < rm$. According to Property 4.2, $lca(lm, v_j) \preceq lca(lm, v_i)$ and $lca(rm, v_i) \preceq lca(rm, v_j)$. According to the definition of closest, $lca(lm, v_i) \prec lca(rm, v_i)$ and $lca(rm, v_j) \preceq lca(lm, v_j)$, which is a contradiction.   □

Property 4.10   Let V and U be lists of nodes, e.g., $V = \{v_1, \cdots, v_l\}$ and $U = \{u_1, \cdots, u_l\}$, such that $V \leq U$, e.g., $v_i \leq u_i$ for $1 \leq i \leq l$. Let $lca(V)$ and $lca(U)$ be the LCA of nodes in V and U, respectively. Then,

1. if $lca(V) \geq lca(U)$, then $lca(U) \preceq lca(V)$,

2. if $lca(V) < lca(U)$, then

   • either $lca(V) \prec lca(U)$,

   • or $lca(V) \nprec lca(U)$, then for any W with $U \leq W$, $lca(V) \nprec lca(W)$.

Proof. This is an extension of Property 4.3 to more than two nodes. The proof is by induction, when V and U contain only two nodes, it is proven in Property 4.3. Assume that it is true for V, U and W, we prove it is true for V', U', W', where $V' = V \cup \{v_l\}$, $U' = U \cup \{u_l\}$, with $v_l \leq u_l$. One important property of lca is that $lca(V') = lca(lca(V), v_l)$. If $lca(U) \preceq lca(V)$, then either $lca(U') \preceq lca(V')$ or $lca(V') \prec lca(U')$. Otherwise, $lca(V) < lca(U)$, according to Property 4.3, there are three cases of $lca(V')$ and $lca(U')$, and we only need to prove the last case, i.e. the case that $lca(V') \nprec lca(U')$. Then for any $W' = W \cup \{w_l\}$, if $lca(U) \preceq lca(W)$, then we are done; otherwise $lca(W) \prec lca(U)$, then $lca(V') \nprec lca(W')$, because $lca(W') \preceq lca(W)$.   □

Table 4.0:

id	k_1	k_2	$\cdots$	k_l
id_m				
$\cdots$				
id_2				
id_1				

Figure 4.3: Stack Data Structure

4.2.2   EFFICIENT ALGORITHMS FOR SLCAS

In this section, we consider three algorithms, namely STACKALGORITHM, INDEXEDLOOKUPEA-GER, and SCANEAGER [Xu and Papakonstantinou, 2005], that find all the $slca(S_1, \cdots, S_l)$ efficiently. Each algorithm has a different characteristic, and it works efficient in some situations. MULTIWAYSLCA further improves the performance of INDEXEDLOOKUPEAGER by proposing some heuristics but with the same worst case time complexity as INDEXEDLOOKUPEAGER. Note that these algorithms only get all the SLCAs, but they do not keep the match nodes for the SLCAs. Finding the match nodes for all the SLCAs can be done efficiently by one scan of SLCAs and one scan of $S_1, \cdots, S_l$, provided that the nodes in SLCAs are in increasing *Dewey ID* order.

Stack Algorithm: This is an adaptation of the stack based sort-merge algorithm [Guo et al., 2003] to compute all the SLCAs. It uses a stack, each stack entry has a pair of components $(id, keyword)$, as shown in Figure 4.3. Assume the id components from the bottom entry to a stack entry en are $id_1, \cdots, id_m$, respectively, then the stack entry en denotes the node with the *Dewey ID* $id_1.id_2.\cdots.id_m$. *keyword* is an array of length l of Boolean values, where $keyword[i] = true$ means that the subtree rooted at the node denoted by the entry contains keyword k_i directly or indirectly.

The general idea of STACKALGORITHM is to use a stack to simulate the postorder traversal of a virtual *XML* tree formed by the union of the paths from root to each node in $S_1, \cdots, S_l$, while the nodes are read in a preorder fashion. When an entry en is popped out, which means that all the descendant-or-self nodes of en in $S_1, \cdots, S_l$ have been visited, it is known whether or not a keyword appears in the subtree. STACKALGORITHM merges all keyword lists and computes the longest common prefix of the node with the smallest *Dewey ID* from the input lists and the node denoted by the top entry of the stack. Then it pops out all top entries until the longest common prefix is reached. If the *keyword* component of a popped entry en contains all the keywords, then the node denoted by en is a SLCA node. Based on the definition of SLCA, all the ancestor nodes of a SLCA node can not be SLCA, so this information is recorded. Otherwise, the keyword containment information of en is used to update its parent entry's *keyword* array. Also, a stack entry is created for each Dewey component of the current visiting node that is not part of the common prefix, where each new entry corresponds to one node on the path from the longest common prefix to the current

---

Algorithm 31 STACKALGORITHM $(S_1, \cdots, S_l)$

Input: l lists of *Dewey IDs*, S_i is the list of *Dewey IDs* of the nodes containing keyword k_i.
Output: All the SLCAs

1: $stack \leftarrow \emptyset$
2: **while** has not reached the end of all Dewey lists **do**
3:      $v \leftarrow getSmallestNode()$
4:      $p \leftarrow lca(stack, v)$
5:      **while** $stack.size > p$ **do**
6:          $en \leftarrow stack.\text{POP}()$
7:          **if** $en.keyword[i] = true, \forall i (1 \leq i \leq l)$ **then**
8:              output en as a SLCA
9:              mark all the entries in $stack$ so that it can never be SLCA node
10:         **else**
11:             $\forall i (1 \leq i \leq l) : stack.\text{TOP}().keyword[i] \leftarrow true,$ if $en.keyword[i] = true$
12:     $\forall i (p < i \leq v.length) : stack.\text{PUSH}(v[i], [])$
13:     $stack.\text{TOP}().keyword[i] \leftarrow true,$ where $v \in S_i$
14: check entries of the stack and return any SLCA node if exists

---

node. Essentially, the node represented by the top entry of the stack is the node visited in pre-order traversal.

STACKALGORITHM is shown in Algorithm 31. It first initializes the stack *stack* to be empty (line 1). As long as there are Dewey lists that have not been visited (line 2), it reads the next node with the smallest *Dewey ID* (line 3), and performs necessary operations. Essentially, reading nodes in this order is equivalent to a preorder traversal of the original *XML* tree ignoring irrelevant nodes. Let $stack[i]$ denote the node represented by the i-th entry of $stack$ starting from the bottom, and $v[i]$ denote the i-th component of the *Dewey ID* of v. After getting v, it computes the LCA of v and the node represented by the top of $stack$ (line 4), which is $stack[p]$. This means that all the keyword nodes have been read that are descendants of $stack[p + 1]$ if they exist, and the keyword containment information has been stored in the corresponding stack entries. Then all those nodes represented by $stack[i]$ ($p < i \leq stack.size$) are popped (lines 5-11). For each popped entry en (line 6), it first checks whether it is a SLCA node (line 7); if en is indeed a SLCA node, then it is output (line 8) and the information is recorded that all its ancestors can not be SLCAs (line 9). Otherwise, the keyword containment information of its parent node is updated (line 11). After popping out all the non-ancestor nodes from $stack$, v and its ancestors are pushed onto $stack$ (line 12), and the keyword containment information is stored (line 13). At this moment, the node represented by the top entry of $stack$ is v, and the whole $stack$ represents all the nodes on the path from root to v, and the keyword containment information is stored compactly. After all the Dewey lists have been read, all the entries need to be popped from *stack*, and a check is performed to see if there exists any SLCA node (line 14).

STACKALGORITHM outputs all the SLCA nodes, i.e. $slca(S_1, \cdots, S_l)$, in time $O(d \sum_{i=1}^{l} |S_i|)$, or $O(ld|S|)$ [Xu and Papakonstantinou, 2005]. Note that the above time complexity does not take into account the time to merge $S_1, \cdots, S_l$, as it will take time $O(d \log l \cdot \sum_{i=1}^{l} |S_i|)$. getSmallestNode (line 3) just retrieves the next node with smallest *Dewey ID* from the merged list.

Indexed Lookup Eager: STACKALGORITHM treats all the Dewey lists $S_1, \cdots, S_l$ equally, but sometimes $|S_1|, \cdots, |S_l|$ vary dramatically. Xu and Papakonstantinou [2005] propose INDEXEDLOOKU-PEAGER to compute all the SLCA nodes, in the situation that $|S_1|$ is much smaller than $|S|$. It is based on the following properties of *slca* function.

Property   4.11 $slca(\{v\}, S) = lca(v, closest(v, S))$,   and $slca(\{v\}, S_2, \cdots, S_l) = slca(slca(\{v\}, S_2, \cdots, S_{l-1}), S_l) = lca(v, closest(v, S_2), \cdots, closest(v, S_l))$ for $l > 2$.

Property 4.11 suggests that we can find the SLCA node of a node, v, and a set of nodes, S, by finding the closest node of v and S first followed by finding the LCA node of v and the closest node of v and S. The definition of *closest* is given in Section 4.1.2. Based on Property 4.11, we can compute $slca(\{v_1\}, S_2, \cdots, S_l)$ by first finding the closest point of v_1 from each set S_i, denoted as $closest(v_1, S_i)$; then finding the *slca* consists of the single node $lca(v_1, closest(v_1, S_2), \cdots, closest(v_1, S_l))$. The computation of $slca(\{v_1\}, S_2, \cdots, S_l)$ takes time $O(d \sum_{i=2}^{l} \log |S_i|)$. Then for arbitrary $S_1, \cdots, S_l$, we have the following property.

Property 4.12 $slca(S_1, \cdots, S_l) = removeAncestor(\bigcup_{v_1 \in S_1} slca(\{v_1\}, S_2, \cdots, S_l))$.

Property 4.12 shows that in order to find SLCA nodes of $S_1, \cdots, S_l$, we can first find $slca(\{v_1\}, S_2, \cdots, S_l)$ for each $v_1 \in S_1$, and then remove all these ancestor nodes. Its correctness follows from the fact that, $slca(S_1, \cdots, S_l) = removeAncestor(lca(S_1, \cdots, S_l))$. The definition of *removeAncestor* is given in Section 4.1.2.

The above two properties directly lead to an algorithm to compute $slca(S_1, \cdots, S_l)$: (1) first compute $\{x_i\} = slca(\{v_i\}, S_2, \cdots, S_l)$, for each $v_i \in S_1$ $(1 \leq i \leq |S_1|)$; (2) $removeAncestor(\{x_1, \cdots, x_{|S_1|}\})$ is the answer. The time complexity of the algorithm is $O(|S_1| \sum_{i=2}^{l} d \log |S_i| + |S_1|d \log |S_1|)$ or $O(|S_1|ld \log |S|)$. The first step of computing $slca(\{v_i\}, S_2, \cdots, S_l)$ for each $v_i \in S_1$ takes time $O(|S_1| \sum_{i=2}^{l} d \log |S_i|)$. The second step takes time $O(|S_1|d \log |S_1|)$, which can be implemented by first sorting $\{x_1, \cdots, x_{|S_1|}\}$ in increasing *Dewey ID* order, and then finding the SLCA nodes by a linear scan. Note that, this time complexity is different from Xu and Papakonstantinou [2005], which is $O(|S_1| \sum_{i=2}^{l} d \log |S_i| + |S_1|^2)$. Although it has the same time complexity of INDEXEDLOOKUPEAGER, the above algorithm is a blocking algorithm, while INDEXEDLOOKUPEAGER is non-blocking.

Lemma 4.13   *Given any two nodes v_i and v_j, with $pre(v_i) < pre(v_j)$, and a set S of Dewey IDs:*

1. *if* $slca(\{v_i\}, S) \geq slca(\{v_j\}, S)$, *then* $slca(\{v_j\}, S) \preceq slca(\{v_i\}, S)$.

2. *if* $slca(\{v_i\}, S) < slca(\{v_j\}, S)$,

   - *either* $slca(\{v_i\}, S)$ *is an ancestor of* $slca(\{v_j\}, S)$,

   - *or* $slca(\{v_i\}, S)$ *is not an ancestor of* $slca(\{v_j\}, S)$, *then for any* v *such that* $pre(v) > pre(v_j)$, $slca(\{v_i\}, S) \not\prec slca(\{v\}, S)$.

The correctness of the above lemma directly follows from Property 4.3 and Property 4.11. It straightforwardly leads to a non-blocking algorithm to compute $slca(S_1, S_2)$, by removing ancestor nodes on-the-fly, which is shown as the subroutine getSLCA in INDEXEDLOOKUPEAGER. The above lemma can be directly applied to multiple sets with the first set as a singleton, i.e. by replacing S by $S_2, \cdots, S_l$ in the lemma. The correctness directly follows Property 4.10, Property 4.9, and Property 4.11.

Property 4.14 $slca(S_1, \cdots, S_l) = slca(slca(S_1, \cdots, S_{l-1}), S_l)$ for $l > 2$.

INDEXEDLOOKUPEAGER, as shown in Algorithm 32, directly follows from Lemma 4.13 and Property 4.11, Property 4.12, and Property 4.14. p in Line 3 is the buffer size, it can be any value ranging from 1 to $|S_1|$; the smaller p is, the faster the algorithm produces the first SLCA. It first computes $X_2 = slca(X_1, S_2)$, where X_1 is the next p nodes from S_1 (line 3). Then it computes $X_3 = slca(X_2, S_3)$ and so on, until it computes $X_l = slca(X_{l-1}, S_l)$ (lines 4-5). Note that at any step, the nodes in X_i are in increasing *Dewey ID* order, and there is no ancestor-descendant relationship between any two nodes in X_i. All nodes in X_l except the first and the last one are guaranteed to be SLCA nodes (line 9). The first node of X_l is checked at line 6. The last node of X_l is carried on to the next iteration (line 9) to be determined whether or not it is a SLCA (line 7). INDEXEDLOOKU-PEAGER outputs all the SLCA nodes, i.e., $slca(S_1, \cdots, S_l)$, in time $O(|S_1| \sum_{i=2}^{l} d \log |S_i|)$, or $O(|S_1| ld \log |S|)$ [Xu and Papakonstantinou, 2005].

Scan Eager: When the keyword frequencies, i.e., $|S_1|, \cdots, |S_l|$, do not differ significantly, the total cost of finding matches by lookups using binary search may exceed the total cost of finding the matches by scanning the keyword lists, i.e $O(|S_1| ld \log |S|) > O(ld|S|)$. SCANEAGER (Algorithm 33) [Xu and Papakonstantinou, 2005] modifies Line 15 of INDEXEDLOOKUPEAGER by using linear scan to find the $lm()$ and $rm()$. It takes advantage of the fact that the accesses to any keyword list are strictly in increasing order in INDEXEDLOOKUPEAGER. Consider the getSLCA(S_1, S_2) subroutine in INDEXEDLOOKUPEAGER, in order to find $lm(v, S_2)$ and $rm(v, S_2)$, SCANEAGER maintains a cursor for each keyword list, and it advances the cursor of S_2 until it finds the node that is closest to v from the left or the right side. Note that if $rm(v, S_2)$ exists, then it should be the next node in S_2 of $lm(v, S_2)$, or the first node in S_2 if $lm(v, S_2) = \bot$. The main idea is based on the fact that, for any v_i and v_j in S_1, with $pre(v_i) < pre(v_j)$, $lm(v_i, S_2) \leq lm(v_j, S_2)$ and $rm(v_i, S_2) \leq rm(v_j, S_2)$, it assumes that all $lm()$ and $rm()$ are not equal to $\bot$. Note that, in order to ensure the correctness of

---

Algorithm 32 INDEXEDLOOKUPEAGER $(S_1, \cdots, S_l)$

Input: l lists of *Dewey IDs*, S_i is the list of *Dewey IDs* of the nodes containing keyword k_i.
Output: All the SLCAs

1: $v \leftarrow \perp$
2: **while** there are more nodes in S_1 **do**
3:     Read p nodes of S_1 into buffer B
4:     **for** $i \leftarrow 2$ **to** l **do**
5:         $B \leftarrow \text{getSLCA}(B, S_i)$
6:     removeFirstNode(B), if $v \neq \perp$ and getFirstNode(B) $\preceq v$
7:     output v as a SLCA, if $v \neq \perp$, $B \neq \emptyset$ and $v \not\prec$ getFirstNode(B)
8:     **if** $B \neq \emptyset$ **then**
9:         $v \leftarrow$ removeLastNode(B)
10:        output B as SLCA nodes
11: output v as a SLCA

12: **Procedure** getSLCA(S_1, S_2)
13: $Result \leftarrow \emptyset; u \leftarrow root$ with *Dewey ID* 0
14: **for** each node $v \in S_1$ in increasing *Dewey ID* order **do**
15:     $x \leftarrow lca(v, closest(v, S_2))$
16:     **if** $pre(u) < pre(x)$ **then**
17:         $Result \leftarrow Result \cup \{u\}$, if $u \not\prec x$
18:         $u \leftarrow x$
19: **return** $Result \cup \{u\}$

SCANEAGER, p at Line (3) must be no smaller than $|S_1|$, i.e., it must first compute $slca(S_1, S_2)$, then $slca(slca(S_1, S_2), S_3)$ and continue. SCANEAGER directly follows from Property 4.10, Property 4.9, Property 4.11.

SCANEAGER outputs all the SLCA nodes, i.e., $slca(S_1, \cdots, S_l)$, in time $O(ld|S_1| + d\sum_{i=2}^{l}|S_i|)$, or $O(ld|S|)$ [Xu and Papakonstantinou, 2005]. Although SCANEAGER has the same time complexity as STACKALGORITHM, it has two advantages. First, SCANEAGER starts from the smallest keyword list, and it does not have to scan to the end of every keyword list and may terminate much earlier. Second, the number of lca operations of SCANEAGER is $O(l|S_1|)$, which is usually much less than that of the STACKALGORITHM that has $O(\sum_{i=1}^{l}|S_i|)$ lca operations.

Multiway SLCA: MULTIWAYSLCA [Sun et al., 2007] further improves the performance of IN-DEXEDLOOKUPEAGER, but with the same worst case time complexity. The Motivation and general idea of MULTIWAYSLCA are shown by the following example.

Algorithm 33 SCANEAGER $(S_1, \cdots, S_l)$

Input: l lists of *Dewey IDs*, S_i is the list of *Dewey IDs* of the nodes containing keyword k_i.
Output: All the SLCAs

1: $u \leftarrow root$ with *Dewey ID* 0
2: **for** each node $v_1 \in S_1$ in increasing *Dewey ID* order **do**
3:      moving cursors in each list S_i to *closest*(v_1, S_i), for $1 \leq i \leq l$
4:      $v \leftarrow lca(v_1, closest(v_1, S_2), \cdots, closest(v_1, S_l))$
5:      **if** $pre(u) < pre(v)$ **then**
6:          **if** $u \not\prec v$ **then**
7:              output u as a SLCA
8:          $u \leftarrow v$
9: output u as a SLCA

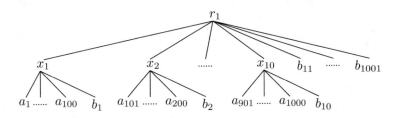

Figure 4.4: An Example *XML* Tree to Illustrate MULTIWAYSLCA [Sun et al., 2007]

Example 4.15   Consider a keyword query $Q = \{a, b\}$ on the *XML* tree shown in Figure 4.4. $S_a = \{a_1, \cdots, a_{1000}\}$ and $S_b = \{b_1, \cdots, b_{1001}\}$, $slca(S_a, S_b) = \{x_1, \cdots, x_{10}\}$. Since $|S_a| < |S_b|$, INDEXEDLOOKUPEAGER will enumerate each of the "a" nodes in S_a in increasing *Dewey ID* order to compute a potential SLCA. This results in a total number of 1000 *slca* computations to produce a result of size 10. Lots of redundant computations have been conducted, e.g., the SLCA of a_i and S_b gives the same result of x_1 for $1 \leq i \leq 100$.

Conceptually, each potential SLCA computed by INDEXEDLOOKUPEAGER can be thought of as being driven by some nodes from S_a (or S_1 in general). But, MULTIWAYSLCA picks an "anchor" node among the l keyword lists to drive the multiway SLCA computation at each individual step. In this example, MULTIWAYSLCA will first consider the first node in each keyword list and select the one with the largest *Dewey ID* as the anchor node. Thus, between $a_1 \in S_a$ and $b_1 \in S_b$, it chooses b_1 as the anchor node. Next, using b_1 as an anchor, it will select the closest node from each other keyword list, i.e., $a_{100} \in S_b$, and will compute the *lca* of those chosen nodes, i.e., $lca(a_{100}, b_1) = x_1$. The next anchor node is selected in the same way by removing all those nodes with *Dewey ID* smaller than $pre(b_1)$ from each keyword list. Then b_2 is selected, and $slca(b_2, S_a) = x_2$. Clearly, MULTIWAYSLCA is able to skip many unnecessary computations.

Definition 4.16 Anchor Node. Given l lists $S_1, \cdots, S_l$, a sequence of nodes, $L = \langle v_1, \cdots, v_l \rangle$ where $v_i \in S_i$, is said to be anchored by a node $v_a \in L$, if for each $v_i \in L$, $v_i = closest(v_a, S_i)$. We refer to v_a as the *anchor node* of L.

Lemma 4.17 *If $lca(L)$ is a SLCA and $v \in L$, then $lca(L) = lca(L')$, where L' is the set of nodes anchored by v in each S_i.*

Thus, it only needs to consider anchored sets, where a set is called anchored if it is anchored by some nodes, for computing potential SLCAs. In fact, from the definition of Compact LCA and its equivalence to ELCA, if a node u is a SLCA, then there must exist a set $\{v_1, \cdots, v_l\}$, where $v_i \in S_i$ for $1 \leq i \leq l$, such that $u = lca(v_1, \cdots, v_l)$ and every v_i is an anchor node.

Lemma 4.18 *Consider two matches $L = \langle v_1, \cdots, v_l \rangle$ and $L' = \langle u_1, \cdots, u_l \rangle$, where $L < L'$, i.e., $v_i \leq u_i$ for $1 \leq i \leq l$, and L is anchored by some node v_i. If L' contains some node u_j with $pre(u_j) \leq pre(v_i)$, then $lca(L')$ is either equal to $lca(L)$ or an ancestor of $lca(L)$.*

Lemma 4.18 provides a useful property to find the next anchor node. Specifically, if we have considered a match L that is anchored by a node v_a, then we can skip all the nodes $v \leq v_a$.

Lemma 4.19 *Let L and L' be two matches. If L' contains two nodes, where one is a descendant of $lca(L)$, while the other is not, then $lca(L') \preceq lca(L)$.*

Lemma 4.19 provides another useful property to optimize the next anchor node. Specifically, if we have considered a match L and $lca(L)$ is guaranteed to be a SLCA, then we can skip all the nodes that are descendants of $lca(L)$.

Lemma 4.20 *Let L be a list of nodes, then $lca(L) = lca(first(L), last(L))$.*

Note that, if the nodes in L is not in order, then $first(L)$ and $last(L)$ will take time $O(ld)$, while directly using the definition also takes time $O(ld)$, i.e., $lca(v_1, \cdots, v_l) = lca(lca(v_1, \cdots, v_{l-1}), v_l)$, where l is the number of nodes in L.

Two algorithms, namely, Basic Multiway-SLCA (BMS) and Incremental Multiway-SLCA (IMS), are proposed in [Sun et al., 2007] to compute all the SLCA nodes. The BMS algorithm implements the general idea above. IMS introduces one further optimization aimed to reduce the *lca* computation of BMS. However, *lca* takes the same time as comparing two *Dewey IDs*, and BMS needs to retrieve nodes in order from an unordered set, and this will incur extra time. So in the

Algorithm 34 MULTIWAYSLCA $(S_1, \cdots, S_l)$

Input: l lists of *Dewey IDs*, S_i is the list of *Dewey IDs* of the nodes containing keyword k_i.
Output: All the SLCAs

1: $v_m \leftarrow last(\{first(S_i) \mid 1 \leq i \leq l\})$, where the index m is also recorded
2: $u \leftarrow root$ with *Dewey ID* 0
3: **while** $v_m \neq \perp$ **do**
4:    **if** $m \neq 1$ **then**
5:       $v_1 \leftarrow closest(v_m, S_1)$
6:       $v_m \leftarrow v_1$, if $v_m < v_1$
7:    $v_i \leftarrow closest(v_m, S_i)$, for each $1 \leq i \leq l, i \neq m$
8:    $x \leftarrow lca(first(v_1, \cdots, v_l), last(v_1, \cdots, v_l))$
9:    **if** $u \leq x$ **then**
10:       output u as a SLCA, if $u \npreceq x$
11:       $u \leftarrow x$
12:    $v_m \leftarrow last(\{rm(v_m, S_i) \mid 1 \leq i \leq l, v_i \leq v_m\})$
13:    **if** $v_m \neq \perp$ **and** $u \npreceq v_m$ **then**
14:       $v_m \leftarrow last(\{v_m\} \cup \{out(u, S_i) \mid 1 \leq i \leq l, i \neq m\})$
15: output u as a SLCA

following, we will show BMS algorithm, denoted as MULTIWAYSLCA, and only show the further optimization of IMS.

MULTIWAYSLCA is shown in Algorithm 34. It computes the SLCAs iteratively. At each iteration, an anchor node v_m is selected to compute the match anchored by v_m and its LCA, where index m is also stored, and v_m is initialized at Line 1. Let u denote the potential SLCA node that is most recently computed, and it is initialized to be the root node with *Dewey ID* 0 (line 2). When v_m is not $\perp$, more potential SLCAs can be found (lines 3-13). Lines 4-6 further optimize the anchor node to be a node with large *Dewey ID* if one exists. After an anchor node v_m is chosen, Line 7 finds the match anchored by v_m, and Line 8 computes the LCA x of this match. If $x \preceq u$ (line 9), then x is ignored. Line 10 outputs u as a SLCA if it is not an ancestor-or-self of x. u is updated to be the recently computed potential SLCA. Lines 12-14 select the next anchor node by choosing the furthest possible node that maximized the number of skipped nodes, where line 12 corresponds to Lemma 4.18, and lines 13-14 corresponds to Lemma 4.19.

Theorem 4.21   *Let u and x be the two variables in* MULTIWAYSLCA. *If $u \geq x$ then $x \preceq u$. Otherwise either $u \prec x$ or $u <\!\!\npreceq x$.*[1] *If $u <\!\!\npreceq x$, then u is guaranteed to be a SLCA.*

---

[1] $u <\!\!\npreceq x$ means that $u < x$ but $u \npreceq x$.

IMS [Sun et al., 2007] further optimizes lines 7-8. Let L denote the match anchored by v_m, i.e., $L = \langle v_1, \cdots, v_l \rangle$. Note that each call of *closest* requires two LCA computations. IMS reduces the number of LCA computation by enumerating all the possible L's whose LCA can be potential SLCA, it can be at most l possible choices. By the definition of match L anchored by v_m, it must satisfy the following three conditions:

- $L \subseteq \{v_m\} \cup P \cup N$, where $P = \{lm(v_m, S_i) \mid i \in [1, l], i \neq m, lm(v_m, S_i) \neq \bot\}$ and $N = \{rm(v_m, S_i) \mid i \in [1, l], i \neq m, rm(v_m, S_i) \neq \bot\}$

- $L \cap S_i \neq \emptyset, \forall i \in [1, l]$

- $v_m \in L$

Without loss of generality, we assume that all $lm(v_m, S_i)$ and $rm(v_m, S_i)$ are not equal to $\bot$, $P = \langle u_1, \cdots, u_{l-1} \rangle$, where $pre(u_i) \leq pre(u_{i+1}) \forall i \in [1, l-2]$, $N = \langle u'_1, \cdots, u'_{l-1} \rangle$ is the list corresponding to P, and $v_m \in S_l$. Then all the possible L's whose LCA can be potential SLCA is of the form $\langle u_i, \cdots, u_{l-1}, v_m, u'_1, \cdots, u'_{i-1} \rangle$, denoted as L_i. This is because that, if $first(L) = u_i$, then $u'_1, \cdots, u'_{i-1}$ must be in L, and L_i is the one with smallest $last(L)$ among all those matches with $first(L) = u_i$, then result in the largest LCA. Note that all the LCAs are on the path from root to v_m, as v_m must be in L. Then we can enumerate L in the order $L_1, \cdots, L_l$, where $first(L_i) \leq first(L_{i+1})$ and $last(L_i) \leq last(L_{i+1})$. Therefore, if $lca(L_i) \preceq lca(L_{i+1}) \forall i < j$, and $lca(L_j) \npreceq lca(L_{j+1})$, then L_j is the match anchored by v_m. Note that, the above discussion are based on the fact that the nodes in P are in increasing *Dewey ID* order, but usually this is not the case, so we have to sort P first.

BMS (MULTIWAYSLCA) and IMS correctly output all the SLCA nodes, i.e. $slca(S_1, \cdots, S_l)$, in time $O(|S_1| l d \log |S|)$ [Sun et al., 2007].

4.3 IDENTIFY MEANINGFUL RETURN INFORMATION

The algorithms shown in the previous section study the efficiency aspect of keyword search. They can find and output all the SLCA nodes (or the whole subtree rooted at SLCA nodes) efficiently. But they do not consider the user's intention for a keyword query. The information returned is either too few (only SLCAs are returned) or too large (the whole subtree rooted at each SLCA is returned). Two approaches have been proposed to identify meaningful return information for a keyword query. One alternative is representing the whole subtree rooted at a SLCA node compactly and presenting it to users, so that it will not overwhelm users [Liu and Chen, 2007]. Another alternative is returning only those subtrees that satisfy two novel properties, which captures desirable *changes* to a query result upon a *change* to the query or data in a general framework [Liu and Chen, 2008b]. Both works are based on the following definition of query result.

Definition 4.22 Keyword Query Results. Processing keyword query Q on *XML* tree T returns a set of query results, denoted as $\mathcal{R}(T, Q)$, where each query result is a subtree (defined by a pair

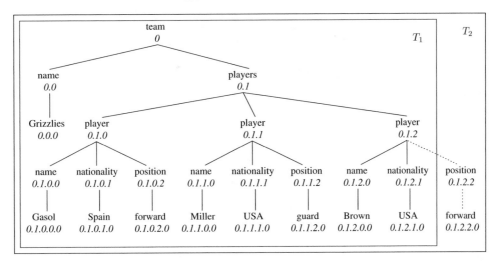

Figure 4.5: Sample *XML* Document [Liu and Chen, 2008b]

$\langle t, M \rangle)$ rooted at t, with nodes (M) corresponding to the matches that are considered relevant to Q. Every keyword in Q has at least one match in M.

Note that one query result should not be subsumed by another; therefore, the root nodes should not have ancestor-descendant relationship, i.e., $t \in slca(Q)$. All the $\langle t, M \rangle$ pairs can be found efficiently, by first finding all SLCAs using the algorithms presented in the previous section, then assigning each match node to the corresponding SLCA node by a linear scan on all SLCAs and $S_1, \cdots, S_l$. In the following, we mainly focus on identifying meaningful information based on $\langle t, M \rangle$.

4.3.1   XSEEK

XSeek [Liu and Chen, 2007; Liu et al., 2009b, 2007] is a system that represents the whole subtree rooted at each SLCA node compactly. We illustrate the general idea of XSeek by the five queries in Figure 4.7 on the *XML* data shown in Figure 4.5. For Q_1, there is only one keyword "Grizzlies;" it is likely that the user is interested in information about "Grizzlies." But by the definition of SLCA, only the node 0.0.0 (Grizzlies) is returned, which is not informative. Ideally, the subtree rooted at 0 (team) should be returned, because this specifies the information that "Grizzlies" is a team name. Consider Q_2 and Q_3, many algorithm will return the same subtree. But, the user is likely to be interested in information about the player whose name is "Gasol" and who is a "forward" in the team for Q_2, and the user is interested in a particular piece of information: the "position" of "Gasol" for Q_3. To process Q_5, XSeek outputs the name of *players* and provides a link to its player children, which provides information about all the players in the team.

<!ELEMENT team (name, players) >
<!ELEMENT name (#PCDATA) >
<!ELEMENT players (play*) >
<!ELEMENT player (name, nationality, position?) >

Figure 4.6: Sample *XML* schema Fragment

Q_1	Grizzlies
Q_2	Gasol, forward
Q_3	Gasol, position
Q_4	team, Grizzlies, forward
Q_5	Grizzlies, players

Figure 4.7: Queries for XSeek

In order to find meaningful return nodes, XSeek analyzes both *XML* data structure and keyword match patterns. Three types of information are represented in *XML* data: entities in the real world, attributes of entities, and connection nodes. The input keywords are categorized into two types: the ones that specify search predicates, and the ones that indicate return information. Then based on the data and keyword analysis, XSeek generates meaningful return nodes.

In order to differentiate the three types of information represented in *XML* data, *XML* schema information is needed, e.g., it is either provided or inferred from the data. An example schema fragment of the *XML* tree shown in Figure 4.5 is shown in Figure 4.6. For each *XML* node, it specifies the names of its sub-elements and attributes using regular expressions with operators * (a set of zero or more elements), + (a set of one or more elements), ? (optional), and | (or). For example, "Element players (player*)" indicates that the "players" can have zero or more "player", "Element player (name, nationality, position ?)" indicates that a "player" should have one "name", one "nationality", and may not have a "position". "Element name (#PCDATA)" specifies that "name" has a value child. In the following, we refer to the nodes that can have siblings of the same name as *-node, as they are followed by "*" in the schema, e.g., the "player" node.

Analyzing *XML* Data Structure: Similar to the *E-R* model used in relational databases, XSeek differentiates nodes in an *XML* tree into three categories.

- A node represents an *entity* if it corresponds to a *-node in the schema.

- A node denotes an *attribute* if it does not correspond to a *-node, and only has one child, which is a value.

- A node is a *connection node* if it represents neither an entity nor an attribute. A connection node can have a child that is an entity, an attribute, or another connection node.

For example, consider the schema shown in Figure 4.6, where "player" is a *-node, indicating a many-to-one relationship with its parent node "players". It is inferred to be an entity, while "name", "nationality", and "position" are considered attributes of a "player" entity. Since "players" is not a *-node and it does not have a value child, therefore, it is considered to be a connection node. Although the above inferences do not always hold, they provide heuristics in the absence of E-R model. When the schema information is not available, it can be inferred based on data summarization [Yu and Jagadish, 2006].

Analyzing Keyword Match Patterns: The input keywords can be classified into two categories: *search predicates*, which correspond to the *where* clause in XQuery or SQL, and *return nodes*, which correspond to the *return* clause in XQuery or *select* clause in SQL. They are inferred as follows,

- If an input keyword k_1 matches a node name (tag) u, and there does not exist an input keyword k_2 matching a node value v, such that u is an ancestor of v, then k_1 specifies a *return node*.

- A keyword that does not indicate a return node is treated as a *predicate* specification. In other words, if a keyword matches a node value, or it matches a node name (tag) that has a value descendant matching another keyword, then this keyword specifies a predicate.

For example, in Q_2 in Figure 4.7, both "Gasol" and "forward" are considered to be *predicates* since they match value nodes. While in Q_3, "position" is inferred as a *return node* since it matches the name of two nodes, neither of which has any descendant value node matching the other keyword.

Generating Search Results: XSeek generates a subtree for each $\langle t, M \rangle$ independently, where $t = lca(M)$ and $t \in slca(Q)$. Sometimes, *return nodes* can be found by analyzing the keyword match patterns, otherwise, they can be inferred implicitly by analyzing the *XML* data and the match M.

Definition 4.23 Master Entity.   If an entity e is the lowest ancestor-or-self of LCA node t of a match M, then e is named the *master entity* of match M. If such an e can not be found, the root of the *XML* tree is considered as the *master entity*.

Based on the previous analysis, we can find the meaningful return information by two steps. First, output all the predicate matches. Second, output the return nodes based on the node category.

Output Predicate Matches: The predicate matches are output, so that the user can check whether the predicates are satisfied in a meaningful way. Therefore, the paths from the LCA node t (or the *master entity* node, if no *return node* found explicitly) to each descendant matches will be output as part of search results, indicating how the keywords are matched and connected to each other.

Output Return Nodes: The *return nodes* are output based their node categories: entity, attribute, and connection node. If it is an attribute node, then its name and value child are output. The subtree rooted at the entity node or connection node is output compactly, by providing the most relevant information at the first stage with expansion links browsing for more details. First, the name of this node and all the attribute children should be output. Then a link is generated to each group of child entities that have the same name (tag), and a link is generated to each child connection node.

For example, for query Q_1, the node 0 (team) is inferred as an implicit *return node*, then the name "team", the names and values of its attributes are output, e.g., 0.0 (name) and 0.0.0 (Grizzlies). An expansion link to its connection child 0.1 (players) is generated.

4.3.2   MAX MATCH

In this work, for each pair $\langle t, M \rangle$, a query result is the tree consisting of the paths in T from t to each match node in M (as well as its value child, if any). The number of query results is denoted as $|R(Q, T)|$. Four properties can be used to prune the irrelevant matches from M [Liu and Chen, 2008b].

Definition 4.24   Delta Result Tree (δ).   Let R be the set of query results of query Q on data T, and R' be the set of updated query results after an insertion to Q or T. A subtree rooted at node v in a query result tree $r' \in R'$ is a *delta result tree* if $desc\text{-}or\text{-}self(v, r') \cap R = \emptyset$ and $desc\text{-}or\text{-}self(parent(v, r'), r') \cap R \neq \emptyset$, where $parent(v, r')$ and $desc\text{-}or\text{-}self(v, r')$ denote the parent and the set of descendant-or-self nodes of v in r', respectively. The set of all delta result trees is denoted as $\delta(R, R')$.

We show the four properties that a query (t, M) should satisfy, namely, *data monotonicity*, *query monotonicity*, *data consistency* and *query consistency*.

Definition 4.25   Data Monotonicity and Data Consistency.   For a query Q and two *XML* documents T and T', $T' = T \cup \{v\}$, where v is an *XML* node not in T.

- An algorithm satisfies *data monotonicity* if the number of query results on T' is no less than that on T. i.e. $|R(Q, T)| \leq |R(Q, T')|$.

- An algorithm satisfies *data consistency* if every delta result tree in $\delta(R(Q, T), R(Q, T'))$ contains v. So there can be either 0 or 1 delta result tree.

Example 4.26   Consider query Q_4 on T_1 and T_2, respectively. Ideally, $R(Q_4, T_1)$ should contain one query result rooted at 0.1.0 (player) with matches 0.1.0.0 (name) and 0.1.0.2.0 (forward). Then consider an insertion of a *position* node with its value *forward* that results in T_2. Ideally, $R(Q_4, T_2)$ should contain one more query result: a subtree rooted at 0.1.2 (player) that matches 0.1.2.0 (name) and 0.1.2.2.0 (forward). Then it will satisfy both *data monotonicity* and *data consistency*, because $|R(Q_4, T_1)| = 1$ and $|R(Q_4, T_2)| = 2$, and the delta result tree is the new result rooted at 0.1.2 (player) which contains the newly added node.

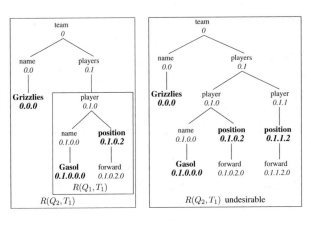

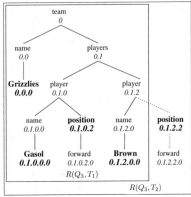

(a) Results of Q_1 and Q_2

(b) Undesirable Results of Q_2 on T_1

(c) Results of Q_3 on T_1 and T_2

Q_1	Gasol, position
Q_2	Grizzlies, Gasol, position
Q_3	Grizzlies, Gasol, Brown, position
Q_4	forward, name
Q_5	forward, USA, name

(d) Sample Queries

Figure 4.8: Sample Queries and Results [Liu and Chen, 2008b]

Consider query Q_3 on T_1 and T_2, respectively, the ideal results are shown in Figure 4.8(c), i.e., $R(Q_3, T_1)$ and $R(Q_3, T_2)$ each contains only one result, and the delta result tree is the subtree rooted at 0.1.2.2 (position) which contains the newly added node.

Definition 4.27   Query Monotonicity and Query Consistency.   For two queries Q and Q' and an *XML* document T, $Q' = Q \cup \{k\}$, where k is a keyword not in Q.

- An algorithm satisfies *query monotonicity* if the number of query results of Q' is no more than that of Q, i.e. $|R(Q, T)| \geq |R(Q', T)|$.

- An algorithm satisfies *query consistency* if every delta result tree in $\delta(R(Q, T), R(Q', T))$ contains at least one match to k.

Example 4.28   Consider queries Q_1 and Q_2 on T_1. Ideally, if $R(Q_1, T_1)$ and $R(Q_2, T_1)$ are as shown in Figure 4.8(a), then they satisfy both *query monotonicity* and *query consistency*, because both queries have one result, and the delta result tree is the subtree rooted at 0.0 (name) which contains the newly added keyword "Grizzlies". While $R(Q_2, T_1)$, as shown in Figure 4.8(b) returned by some algorithms violate *query consistency*. Compared with $R(Q_1, T_1)$ as shown in Figure 4.8(a), the delta result tree contains two subtrees, one is the subtree rooted at 0.0 (name) which contains "Grizzlies", and the other is rooted at 0.1.1 (player) which does not contain "Grizzlies".

Consider query Q_4 and Q_5 on T_2. Ideally, $R(Q_4, T_2)$ will contain two subtrees, one is rooted at 0.1.0 (player) and the other is rooted at 0.1.2 (player), while $R(Q_5, T_2)$ will contain only one subtree rooted 0.1.2 (player) with matches 0.1.2.0 (name), 0.1.2.1.0 (USA) and 0.1.2.2.0 (forward). Then it will satisfy both *query monotonicity*, i.e., $R(Q_4, T_2) = 2$ and $R(Q_5, T_2) = 1$, and *query consistency*, i.e., the delta result tree is the subtree rooted 0.1.2.1 (nationality) which contains the newly added keyword "USA".

Max Match Algorithm: MAXMATCH algorithm [Liu and Chen, 2008b] is proposed to find relevant subtrees that satisfies these four properties. Recall that the result is defined as $r = (t, M)$, where $t \in slca(Q)$ is a SLCA and M are match nodes. Actually, there is one result for each $t \in slca(Q)$. So in the following we will show how to find relevant matches M among all the matches nodes that are descendant of t, guided by the four properties.

Definition 4.29   Descendant Matches.   For a query Q on *XML* data T, the descendant matches of a node v in T, denoted as $dMatch(v)$, is the set of keywords in Q that appear in the subtree rooted at v in T.

Definition 4.30   Contributor.   For a query Q on *XML* data T, a node v in T is called a *contributor* to Q, if (1) v has an ancestor-or-self $v_1 \in slca(Q)$, and (2) v does not have a sibling v_2, such that $dMatch(v) \subset dMatch(v_2)$.

Consider query Q_2 on the *XML* document T_1, $dMatch(0.1.0) = \{Gasol, position\}$, and $dMatch(0.1.1) = \{position\}$. $dMatch(0.1.1) \subset dMatch(0.1.0)$; therefore, node 0.1.1 (player) is not a contributor.

Definition 4.31   Relevant Match.   For an *XML* tree T and a query Q, a match node v in T is relevant to Q, if (1) v has an ancestor-or-self $u \in slca(Q)$, and (2) every node on the path from u to v is a contributor to Q.

Algorithm 35 MaxMatch $(S_1, \cdots, S_l)$

Input: l lists of *Dewey IDs*, S_i is the list of *Dewey IDs* of the nodes containing keyword k_i.
Output: All the SLCA nodes t together with its relevant subtree

1:   $SLCAs \leftarrow slca(S_1, \cdots, S_l)$
2:   $group \leftarrow \text{groupMatches}(SLCA, S_1, \cdots, S_l)$
3:   **for** group $(t, M) \in group$ **do**
4:     pruneMatches(t, M)

5:   **Procedure** pruneMatches(t, M)
6:   **for** $i \leftarrow 1$ **to** $M.size$ **do**
7:     $u \leftarrow lca(M[i], M[i + 1])$
8:     **for** each node v on the path from $M[i]$ to u (exclude u) **do**
9:       $v.dMatch[j] \leftarrow true$, if v contains keyword k_j
10:      let v_p and v_c denote the parent and child of v on this path
11:      $v.dMatch \leftarrow v.dMatch$ **OR** $v_c.dMatch$
12:      $v.last \leftarrow i$
13:      $v_p.dMatchSet[num(v.dMatch)] \leftarrow true$
14:   $i \leftarrow 1; u \leftarrow t$; output t
15:   **while** $i \leq M.size$ **do**
16:     **for** each node v from u (exclude u) to $M[i]$ **do**
17:      **if** isContributor(v) **then**
18:       output v
19:      **else**
20:       $i \leftarrow v.last$; **break**
21:     $i \leftarrow i + 1; u \leftarrow lca(M[i - 1], M[i])$

Continue the query Q_2 on T_1, the node 0.1.1 (player) is not a contributor, then match node 0.1.1.2 (position) is irrelevant to Q. So the subtree shown in Figure 4.8(b) can not be returned, in order to satisfy the four properties.

Definition 4.32 Query Results of MaxMatch **.** For an *XML* tree T and a query Q, each query result generated by MaxMatch is defined by $r = (t, M), \forall t \in slca(Q)$, where M is the set of relevant matches to Q in the subtree rooted at t.

The subtree shown in Figure 4.8(b) will not be generated by MaxMatch, because 0.1.1.2 (position) is not a relevant match, and because 0.1.1 is not a contributor. Note that there exists exactly one tree returned by MaxMatch for each $t \in slca(Q)$.

MaxMatch is shown in Algorithm 35. It consists of three steps: computing SLCAs, group-Matches, and pruneMatches. In the first step (line 1), it computes all the SLCAs. It can use any of the previous algorithms, and we will use STACKALGORITHM or SCANEAGER, which takes time $O(d \sum_{i=1}^{l} |S_i|)$, or $O(ld|S|)$. However, groupMatches needs to do a *Dewey ID* comparison for each match, pruneMatches needs to do both a postorder and a preorder traversal of the match nodes, which subsume the time complexity of $O(d \sum_{i=1}^{l} |S_i|)$.

In the second step (line 2), groupMatches groups the matched nodes in $S_1, \cdots, S_l$ to each SLCA node computed in the first step. This can be implemented by first merging $S_1, \cdots, S_l$ into a single list in increasing *Dewey ID* order, then adding the match nodes to the corresponding SLCA node with $O(d)$ amortized time (because at least one *Dewey ID* comparison is needed). The algorithm is based on the fact that, (1) each match can be a descendant of at most one SLCA, (2) if $t_1 < t_2$, then all the descendants of t_1 precede all the descendants of t_2. groupMatches takes $O(d \log l \sum_{i=1}^{l} |S_i|)$ time, which is the time to merge l sorted lists $S_1, \cdots, S_l$. Note that Liu and Chen [2008b] analyze the time of merge as $O(\log l \sum_{i=1}^{l} |S_i|)$ based on the assumption that comparing two match nodes takes $O(1)$ time. It takes $O(d)$ time if only *Dewey ID* is presented.

In the third step (line 3), pruneMatches computes relevant matches for each SLCA t, with M storing all the descendant match nodes. It consists of both a postorder and a preorder traversal of the subtree which is a union of all the paths from t to each match node in M. Lines 6-13 conduct the postorder traversal, during which it finds the *descendant matches* for each node, stored in $v.dMatch$, which is a Boolean array of size l (and can be compactly represented by *int* values where each *int* value represents 32 (or 64) elements of Boolean array). $v.dMatchSet$ stores the information of all the possible descendant matches its children have, which is used to determine whether a node is a contributor or a node (line 17). $v.last$ stores the index of the last descendant nodes of v, which is used to skip to the next match node that might be relevant (line 20). Lines 14-21 conduct the preorder traversal. For each node v visited (line 16), if it is a contributor, then it is output, otherwise all the descendant match nodes of v can not be relevant, and the algorithm skips to the next match node that is not a descendant of v (line 20). isContributor can be implemented in different ways. One is iterating over all of $dMatch$'s siblings to check whether there is a sibling that contains superset keywords. The other is iterating over $dMatchSet$ (which is of size 2^l) [Liu and Chen, 2008b] that works better when l is very small and the fan-out of nodes is very large (i.e., greater than 2^l).

Theorem 4.33   *[Liu and Chen, 2008b] The subtrees generated by* MAXMATCH *satisfies all four properties, namely, data monotonicity, data consistency, query monotonicity and query consistency, and* MAXMATCH *will generate exactly one subtree rooted at each node* $t \in slca(Q)$.

4.4   ELCA-BASED SEMANTICS

ELCAs is a superset of SLCAs, and it can find some relevant information that SLCA can not find, e.g., in Figure 4.1, node 0 (school) is an ELCA for keyword query $Q = \{John, Ben\}$, which captures the information that "Ben" participates in a sports club in the school that "John" is the dean. In this section, we show efficient algorithms to compute all ELCAs and properties to capture relevant subtrees rooted at each ELCA.

Algorithm 36 DEWEYINVERTEDLIST $(S_1, \cdots, S_l)$

Input: l list of *Dewey IDs*, S_i is the list of *Dewey IDs* of the nodes containing keyword k_i.
Output: All the ELCA nodes

1: $stack \leftarrow \emptyset$
2: **while** has not reached the end of all Dewey lists **do**
3:     $v \leftarrow$ getSmallestNode()
4:     $p \leftarrow lca(stack, v)$
5:     **while** $stack.size > p$ **do**
6:        $en \leftarrow stack.\text{POP}()$
7:        **if** $en.keyword[i] = true, \forall i\,(1 \leq i \leq l)$ **then**
8:          output en as a ELCA
9:          $en.ContainsAll \leftarrow true$
10:        **else if not** $en.ContainsAll$ **then**
11:          $\forall i\,(1 \leq i \leq l) : stack.\text{TOP}().keyword[i] \leftarrow true,$ if $en.keyword[i] = true$
12:        $stak.\text{TOP}().ContainsAll \leftarrow true,$ if $en.ContansAll$
13:     $\forall i\,(p < i \leq v.length) : stack.\text{PUSH}(v[i], [])$
14:     $stack.\text{TOP}().keyword[i] \leftarrow true,$ where $v \in S_i$
15: check entries of the stack and return any ELCA if exists

4.4.1 EFFICIENT ALGORITHMS FOR ELCAS

ELCA-based semantics for keyword search is first proposed by Guo et al. [2003], who also propose ranking functions to rank trees. In their ranking method, there is an *Elem Rank* value for each node, which is computed similar to PageRank [Brin and Page, 1998], working on the graph formed by considering hyperlink edges in *XML*. The score of a subtree is a function of the decayed *Elem Rank* value of match nodes by the distance to the root of the subtree. An adaptation of Threshold Algorithm [Fagin et al., 2001] is used to find the top-K subtrees. However, there is no guarantee on the efficiency, and it may perform worse in some situations.

Dewey Inverted List: DEWEYINVERTEDLIST (Algorithm 36) [Guo et al., 2003] is a stack based algorithm, and it works by a postorder traversal on the tree formed by the paths from root to all the match nodes. The general idea of this algorithm is the same as STACKALGORITHM, and actually STACKALGORITHM is an adaptation of DEWEYINVERTEDLIST to compute all the SLCAs.

    DEWEYINVERTEDLIST is shown in Algorithm 36. It reads match nodes in a preorder traversal (line 3), using a stack to simulate the postorder traversal. When a node en is popped out from *stack*, all its descendant nodes have been visited, and the keyword containment information is stored in *keyword* component of *stack*. If the keyword component of en is true for all entries, then en is an ELCA, and $en.ContainsAll$ is set to *true* to record this information. $en.ContainsAll$ means that the subtree rooted at en contains all the keywords, then its keyword containment information should not be updated to its parent node (line 10), but it still can be an ELCA node if it contains all the keywords in other paths (line 7).

DEWEYINVERTEDLIST outputs all the ELCA nodes, i.e., $elca(S_1, \cdots, S_l)$, in time $O(d \sum_{i=1}^{l} |S_i|)$, or $O(ld|S|)$, where the time to merge l ordered list $S_1, \cdots, S_l$ is not included [Guo et al., 2003].

Indexed Stack: The INDEXEDSTACK algorithm is based on the following property, where the correctness is guaranteed by the definition of Compact LCA and its equivalence to ELCA, i.e., a node $u = lca(v_1, \cdots, v_l)$ is a CLCA with respect to $v_1, \cdots, v_l$, if and only if u dominates each v_i, i.e., $u = slca(S_1, \cdots, S_{i-1}, v_i, S_{i+1}, \cdots, S_l)$.

Property 4.34 $elca(S_1, \cdots, S_l) \subseteq \bigcup_{v_1 \in S_1} slca(\{v_1\}, S_2, \cdots, S_l)$

Let $elca\_can(v_1)$ denote $slca(\{v_1\}, S_2, \cdots, S_l)$, and $elca\_can(S_1, \cdots, S_l)$ denote $\cup_{v_1 \in S_1} elca\_can(v_1)$. The above property says that $elca\_can(S_1, \cdots, S_l)$ is a candidate ELCA that is a superset of the ELCAs. We call a node v an ELCA_CAN if $v \in elca\_can(S_1, \cdots, S_l)$. Based on the above property, the algorithm to find all the ELCAs can be decomposed into two step: (1) first find all ELCA_CANs, (2) then find ELCAs in ELCA_CANs. ELCA_CANs can be found by INDEXEDLOOKUPEAGER in time $O(|S_1| \sum_{i=2}^{l} d \log |S_i|)$, or $O(|S_1| ld \log |S|)$. In the following, we mainly focus on the second step (function isELCA), which checks whether v is an ELCA for each $v \in elca\_can(S_1, \cdots, S_l)$.

Function isELCA: Let $child\_elcacan(v)$ denote the set of children of v that contain all the l keywords. Equivalently, $child\_elcacan(v)$ is the set of child nodes u of v such that either u or one of u's descendant nodes is an ELCA_CAN, i.e.

$$child\_elcacan(v) = \{u \in child(v) \mid \exists x \in elca\_can(S_1, \cdots, S_l), u \preceq x\}$$

where $child(v)$ is the set of children of v. Assume $child\_elcacan(v)$ is $\{u_1, \cdots, u_m\}$ as shown in Figure 4.9. According to the definition of ELCA, a node v is an ELCA if and only if it has ELCA witness nodes $n_1 \in S_1, \cdots, n_l \in S_l$, and each n_i is not in any subtree rooted at the nodes from $child\_elcacan(v)$.

To determine whether v is an ELCA or not, we probe every S_i to see if there is a node $x_i \in S_i$ such that x_i is (1) either in the forest under v to the left of the path vu_1, i.e., in the *Dewey ID* range $[pre(v), pre(u_1))$; (2) or in any forest F_{i+1} that is under v and between the paths vu_i and vu_{i+1}, for $1 \leq i < m$, i.e., in the *Dewey ID* range $[p.(c+1), pre(u_{i+1}))$, where $p.c$ is the *Dewey ID* of u_i, then $p.(c+1)$ is the *Dewey ID* for the immediate next sibling of u_i; (3) or in the forest under v to the right of the path vu_m. Each case can be checked by a binary search on S_i. The procedure isELCA [Xu and Papakonstantinou, 2008] is shown in Algorithm 37, where ch is the list of nodes in $child\_elcacan(v)$ in increasing *Dewey ID* order. Line 3-8 check the first and the second case, and lines 9-10 check the last case. The time complexity of isELCA is $O(|child\_elca(v)|ld \log |S|)$.

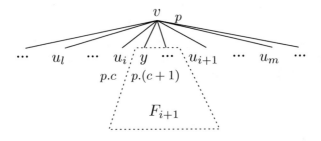

Figure 4.9: v and *child_elcacan(v)* [Xu and Papakonstantinou, 2008]

Algorithm 37 ISELCA (v, ch)

Input: a node v and $ch = child\_elcacan(v)$.
Output: return *true* if v is ELCA, *false* otherwise

1: **for** $i \leftarrow 1$ **to** l **do**
2:     $x \leftarrow v$
3:     **for** $j \leftarrow 1$ **to** $|ch|$ **do**
4:         $x \leftarrow rm(x, S_i)$
5:         **if** $x = \bot$ **or** $pre(x) < pre(ch[j])$ **then**
6:            **break**
7:         **else**
8:            $x \leftarrow \text{nextSibling}(ch[j])$
9:     **if** $j = |ch| + 1$ **then**
10:         **return** *false*, if $v \not\prec rm(x, S_i)$
11: **return** *true*

After the first step that we got ELCA_CANs, if we can find *child_elcacan(v)* efficiently for each ELCA_CAN v, then we can find ELCAs in time $O(|S_1| l d \log |S|)$. If we assign each ELCA_CAN u to be the child of its ancestor ELCA_CAN node v with the largest *Dewey ID*, then u corresponds to exactly one node in *child_elcacan(v)*, and the node in *child_elcacan(v)* corresponding to u can be found in $O(d)$ time by the *Dewey ID*. In the following, we use *child_elcacan(v)* to denote the set of ELCA_CAN nodes u which is a descendant of v and there does not exist any node x with $v \prec x \prec w$, i.e.

$$child\_elcacan(v) = \{u \in elca\_can(S_1, \cdots, S_l) \mid v \prec u \wedge$$
$$\nexists x \in elca\_can(S_1, \cdots, S_l)(v \prec x \prec u)\}$$

There is an one-to-one correspondence between the two definitions of *child_elcacan(v)*. It is easy to see that $\sum_{v \in elca\_can(S_1, \cdots, S_l)} |child\_elcacan(v)| = O(|elca\_can(S_1, \cdots, S_l)|) = O(|S_1|)$.

Now the problem becomes how to compute $child\_elcacan(v)$ efficiently for all $v \in elca\_can(S_1, \cdots, S_l)$. Note that, the nodes in $elca\_can(S_1, \cdots, S_l)$ as computed by $\cup_{v_1 \in S_1} elca\_can(v_1)$ are not sorted in *Dewey ID* order. Similar to DEWEYINVERTEDLIST, a stack based algorithm is used to compute $child\_elcacan(v)$, but it works on the set $elca\_can(S_1, \cdots, S_l)$, while DEWEYINVERTEDLIST works on the set $S_1 \cup S_2 \cdots \cup S_l$. Each stack entry created for a node $v_1 \in S_1$ has the following three components:

- $elca\_can$ is $elca\_can(v_1)$

- CH is $child\_elcacan(v_1)$

- SIB is the list of ELCA_CANs before $elca\_can$, which is used to compute CH

INDEXEDSTACK [Xu and Papakonstantinou, 2007, 2008] is shown in Algorithm 38. For each node $v_1 \in S_1$, it computes $elca\_can_{v_1} = elca\_can(v_1)$ (line 3), a stack entry en is created for $elca\_can_{v_1}$ (line 4). If the *stack* is empty (line 5), we simply push en to *stack* (line 6). Otherwise, different operations are applied based on the relationship between $elca\_can_{v_1}$ and $elca\_can_{v_2}$, which is the node at the top of *stack*.

- $elca\_can_{v_1} = elca\_can_{v_2}$, then en is discarded (lines 8-9)

- $elca\_can_{v_2} \prec elca\_can_{v_1}$, then just push en to *stack* (lines 10-11),

- $elca\_can_{v_2} < elca\_can_{v_1}$, but $elca\_can_{v_2} \nprec elca\_can_{v_1}$, then the non-ancestor nodes of $elca\_can_{v_1}$ in *stack* is popped out, and it is checked whether it is an ELCA or not (procedure popStack (lines 23-30)), because all its descendant match nodes have been read, and the $child\_elcacan$ information have been stored in $popEntry.CH$ (lines 27-28). After the non-ancestor nodes have been popped out (line 13), it may be necessary to store the sibling nodes of en to $en.SIB$. Note that, in this case, there may exist a potential ELCA that is the ancestor of en, and the descendant of the top entry of the *stack* (or the root of the *XML* tree if *stack* is empty). If this is possible (line 15), then the sibling information is stored in $en.SIB$ (line 16).

- $elca\_can_{v_1} \prec elca\_can_{v_1}$, then the non-ancestor nodes of $elca\_can_{v_1}$ in *stack* is popped out, and it is checked whether it is to be an ELCA or not (line 19), and $en.CH$ is stored (line 20). Note that there does not exist any more potential ELCA nodes that are descendants of the popped entries.

Note that these are the only four possible cases of the relationship between $elca\_can_{v_1}$ and $elca\_can_{v_2}$. INDEXEDSTACK output all the ELCA nodes, i.e., $elca(S_1, \cdots, S_l)$, in time $O(|S_1| \sum_{i=2}^{l} d \log |S_i|)$, or $O(|S_1| l d \log |S|)$ [Xu and Papakonstantinou, 2008].

Algorithm 38 INDEXEDSTACK $(S_1, \cdots, S_l)$

Input: l list of *Dewey IDs*, S_i is the list of *Dewey IDs* of the nodes containing keyword k_i.
Output: output all ELCAs

1: $stack \leftarrow \emptyset$
2: **for** each node $v_1 \in S_1$, in increasing *Dewey ID* order **do**
3:   $elca\_can_{v_1} \leftarrow slca(\{v_1\}, S_2, \cdots, S_l)$
4:   $en \leftarrow [elca\_can \leftarrow elca\_can_{v_1}; SIB \leftarrow []; CH \leftarrow []]$
5:   **if** $stack = \emptyset$ **then**
6:     $stack.\text{PUSH}(en);$ **continue**
7:   $topEntry \leftarrow stack.\text{TOP}(); elca\_can_{v_2} \leftarrow topEntry.elca\_can$
8:   **if** $elca\_can_{v_1} = elca\_can{v_2}$ **then**
9:     $\perp$
10:   **else if** $elca\_can_{v_2} \prec elca\_can_{v_1}$ **then**
11:     $stack.\text{PUSH}(en)$
12:   **else if** $elca\_can_{v_2} < elca\_can_{v_1}$ **then**
13:     $popEntry \leftarrow popStack(elca\_can_{v_1})$
14:     $top\_elcacan \leftarrow stack.\text{TOP}().elca\_can$
15:     **if** $stack \neq \emptyset$ **and** $top\_elcacan \prec lca(elca\_can_{v_1}, popEntry.elca\_can)$ **then**
16:       $en.SIB \leftarrow [popEntry.SIB, popEntry.elca\_can]$
17:     $stack.\text{PUSH}(en)$
18:   **else if** $elca\_can_{v_1} \prec elca\_can_{v_2}$ **then**
19:     $popEntry \leftarrow popStack(elca\_can_{v_1})$
20:     $en.CH \leftarrow [popEntry.SIB, popEntry.elca\_can]$
21:     $stack.\text{PUSH}(en)$
22: $popStack(0)$

23: **Procedure** popStack $(elca\_can_{v_1})$
24: $popEntry \leftarrow \perp$
25: **while** $stack \neq \emptyset$ **and** $stack.\text{TOP}().elca\_can \nprec elca\_can_{v_1}$ **do**
26:   $popEntry \leftarrow stack.\text{POP}()$
27:   **if** isELCA $(popEntry.elca\_can, toChild\_elcacan(popEntry.elca\_can, popEntry.CH))$ **then**
28:     output $popEntry.elca\_can$ as an ELCA
29:   $stack.\text{TOP}().CH \leftarrow stack.\text{TOP}().CH + popEntry.elca\_can$
30: **return** $popEntry$

4.4.2   IDENTIFYING MEANINGFUL ELCAS

Kong et al. [2009] extend the definition of *contributor* [Liu and Chen, 2008b] to *valid-contribute*, and they propose an algorithm similar to MAXMATCH to compute relevant matches based on ELCA semantics, i.e., root t can be an ELCA node.

Definition 4.35   Valid Contributor.   Given an *XML* data T and a keyword query Q, a node v in Q is called a *valid contributor* to Q, if either one of the following two conditions holds:

  1. v has a unique label $tag(v)$ among its sibling nodes

2. v has several siblings $v_1, \cdots, v_m (m \geq 1)$, with the same label as $tag(v)$, but the following conditions hold:

- $\nexists v_i, dMatch(v) \subset dMatch(v_i)$

- $\forall v_i > v$, if $dMatch(v) = dMatch(v_i)$, then $TC_v \neq TC_{v_i}$, where TC_v denote the set of words (among the match nodes in M) appear in the subtree rooted at v

A *valid contributor* only compares nodes with its sibling nodes that have the same label. If a node v has a unique label among its sibling nodes, then it is a valid contributor. Otherwise, only those nodes whose $dMatch$ is not subsumed by any sibling node with the same label is a valid contributor. Also, if the subtree rooted at two sibling nodes contains exactly the same set of words (TC_v), then only one is a valid contributor.

Definition 4.36   Relevant Match.   For an *XML* tree T and a query Q, a match node v in T is relevant if v is a witness node of $u \in elca(Q)$, and all the nodes on the path from u to v are valid contributors.

Based on this definition of *valid contributor* and *relevant match*, all the subtrees formed by ELCA node and its corresponding relevant match nodes will satisfy the four properties discussed earlier [Liu and Chen, 2008b], namely, *data monotonicity*, *data consistency*, *query monotonicity*, and *query consistency*. An algorithm to find the relevant matches for each ELCA node exists [Kong et al., 2009], that consists of three steps: (1) find all ELCAs using DEWEYINVERTEDLIST or INDEXED-STACK, (2) group match nodes to each ELCA node, (3) prune irrelevant matches from each group. The algorithm uses ideas similar to MAXMATCH to find relevant matches according to the definition of valid contributor.

4.5   OTHER APPROACHES

There exist several semantics other than SLCA and ELCA for keyword search on *XML* databases, namely, meaningful LCA (MLCA) [Li et al., 2004, 2008b], *interconnection* [Cohen et al., 2003], *Compact Valuable LCA (CVLCA)* [Li et al., 2007a], and relevance oriented ranking [Bao et al., 2009]. The difference between MLCA and *interconnection* is that MLCA is based on SLCA, whereas *interconnection* is not, i.e., the root nodes of the subtrees returned by *interconnection* may not be a SLCA node. CVLCA is a combination of ELCA semantics and the *interconnection* semantics.

Another approach to keyword search on *XML* databases is to make use of the schema information where results are minimal connected trees of *XML* fragments that contain all the keywords [Balmin et al., 2003; Hristidis et al., 2003b]. Hristidis et al. study keyword search on *XML* trees, and propose efficient algorithms to find minimum connecting trees [Hristidis et al., 2006]. Al-Khalifa et al. integrate the IR-styled ranking function into XQuery, and they propose a bulk-algebra which is the basis for integrating information retrieval techniques into a standard pipelined database query evaluation engine [Al-Khalifa et al., 2003]. NaLIX (Natural Language Interface to XML) is a system, in which an arbitrary English language sentence is translated into an XQuery expression, and it can be evaluated against an *XML* database [Li et al., 2007b]. The problem of keyword search on *XML* using a minimal number of materialized views is also studied, where the answer definition is based on SLCA semantics [Liu and Chen, 2008a]. Some works study the problem of keyword search over virtual (unmaterialized) *XML* views [Shao et al., 2007, 2009a]. eXtract is a system to generate snippets for tree results of querying on *XML* database, which highlights the most dominant features [Huang et al., 2008a,b]. Answer differentiation is studied to find a limited number of valid features in result so that they can maximally differentiate this result from the others [Liu et al., 2009a].

Other Topics for Keyword Search on Databases

In this chapter, we discuss several interesting research issues regarding keyword search on databases. In Section 5.1, we discuss some approaches that are proposed to select some *RDB* among many to answer a keyword query. In Section 5.2, we discuss keyword search in a spatial database. In Section 5.3, we introduce a PageRank based approach called *ObjectRank* in *RDB*, and an approach that projects a database that only contains tuples relating to a keyword query.

5.1   KEYWORD SEARCH ACROSS DATABASES

There are two main issues to be considered in keyword search across multiple databases:

1. When the number of databases is large, a proper subset of databases need to be selected that are most suitable to answer a keyword query. This is the problem of keyword-based selection of the top-k databases, and it is studied in *M-KS* [Yu et al., 2007] and *G-KS* [Vu et al., 2008].

2. The keyword query needs to be executed across the databases that are selected. This problem is studied in *Kite* [Sayyadian et al., 2007].

5.1.1   SELECTION OF DATABASES

In order to rank a set of databases $\mathcal{D} = \{D_1, D_2, \cdots\}$ according to the their suitability to answer a certain keyword query Q, a score function $score(D, Q)$ is defined for each database $D \in \mathcal{D}$. In the ideal case, if the keyword query is evaluated in each database individually, the best database to answer the query is the one that can generate high quality results. Suppose $\mathcal{T} = \{T_1, T_2, ...\}$ is the set of results (*MTJNT*s, see Chapter 2) for query Q over database D. The following equation can be used to score database D:

$$score(D, Q) = \sum_{T \in \mathcal{T}} score(T, Q) \qquad (5.1)$$

where $score(T, Q)$ can be any scoring function for the *MTJNT* T as discussed in Chapter 2.

In practice, it is inefficient to evaluate Q on every database $D \in \mathcal{D}$. A straightforward way to solve the problem efficiently is to calculate the keyword statistics for each $k_i \in Q$ on each database $D \in \mathcal{D}$ and summarize the statistics as a score reflecting the relevance of Q to D. There are two drawbacks to this solution. First, the keyword statistics can not reveal the importance of the keyword to the databases. For example, a term in a primary key attribute of a table may be referred to by a large number of foreign keys. Such a term may be very important in answering the keyword query, but its frequency in the database can be very low. Furthermore, two different keywords can be connected through a sequence of foreign key references in a relational database. The length and number of such connections may largely reveal the capability of the database to answer a certain keyword query. The statistics of single keywords can not capture such relationships between keywords, and thus they may choose a database that has high keyword frequency, but they may not generate any $MTJNT$.

Suppose the keyword space is $\mathcal{K} = \{w_1, w_2, ..., w_s\}$. For each database $D \in \mathcal{D}$, we can construct a keyword relationship matrix (KRM) $\mathcal{R} = (r_{i,j})_{s \times s}$, which is a s by s matrix where each element is defined as follows:

$$\mathcal{R}_{i,j} = \sum_{d=0}^{\delta} \varphi_d \cdot \omega_d(w_i, w_j) \tag{5.2}$$

Here, $\omega_d(w_i, w_j)$ is the number of joining sequences of length d: $t_0 \bowtie t_1 \bowtie ... \bowtie t_d$ where $t_i \in D$ $(1 \leq i \leq d)$ is a tuple, and t_0 contains keyword w_i and t_d contains keyword w_j. δ is a parameter to control the maximum length of the joining sequences, because it is meaningless if two tuples t_0 and t_j are too far away from each other. φ_d is a function of d that measures the importance of the joining sequence of length d, it can be specified based on different requirements.

$$\varphi_d = \frac{1}{d+1} \tag{5.3}$$

For example, in $M\text{-}KS$ [Yu et al., 2007], the value $\omega_d(w_i, w_j)$ increases exponentially with respect to d, so another control parameter M is set such that if the value $\sum_{d=0}^{\delta} \omega_d(w_i, w_j) > M$, the $\mathcal{R}_{i,j}$ value is changed to be:

$$\mathcal{R}_{i,j} = \sum_{d=0}^{\delta'-1} \varphi_d \cdot \omega_d(w_i, w_j) + \varphi_{\delta'} \cdot (M - \sum_{d=0}^{\delta'-1} \omega_d(w_i, w_j)) \tag{5.4}$$

where δ' is a value such that $\delta' \leq \delta$ and $\sum_{d=0}^{\delta'} \omega_d(w_i, w_j) \geq M$ and $\sum_{d=0}^{\delta'-1} \omega_d(w_i, w_j) < M$, i.e., $\delta' = min\{\delta_p | \sum_{d=0}^{\delta_p} \omega_d(w_i, w_j) \geq M\}$.

Given the KRM of database D, and a keyword query Q, the $score(D, Q)$ can be calculated as follows.

$$score(D, Q) = \sum_{w_i \in Q, w_j \in Q, i < j} \mathcal{R}_{i,j} \tag{5.5}$$

In place of summation, it is possible to use aggregate functions $min, max,$ or $product$ according to different requirements.

A number of drawbacks of KRM have been identified [Vu et al., 2008]. First, KRM only considers the pairwise relationship between keywords in a query, and this may generate many false

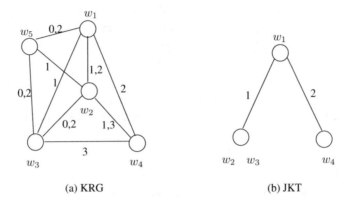

(a) KRG   (b) JKT

Figure 5.1: *KRG* and one of its *JKT* for query $Q = \{w_1, w_2, w_3, w_4\}$

positives because each real result *MTJNT* constructs all keywords in the shape of a tree rather than a pairwise graph. Second, considering only the connections between keywords in a relational database is not enough to rank databases; it is important to also integrate IR-Styled score in the scoring function. These can be addressed as follows.

Suppose the keyword space is $\mathcal{K} = \{w_1, w_2, ...\}$. For each database $D \in \mathcal{D}$, a keyword relationship graph (*KRG*) can be constructed, $G(V, E)$, where, for each keyword $w_i \in Q$, there is a node $w_i \in V(G)$, and for every two keywords $w_i \in Q$ and $w_j \in Q$, if w_i and w_j can be connected through at least one joining sequence of tuples in D, then an edge $(w_i, w_j) \in E(G)$ is added. For each edge $(w_i, w_j) \in E(G)$, a set of weights are assigned. More precisely, when there is a joining sequence of tuples with length d that connect w_i and w_j in the two ends, then a weight d is added to the edge (w_i, w_j) in G.

Given the *KRG* G for database D, and a keyword query $Q = \{k_1, k_2, ..., k_l\}$, a Join Keyword Tree (*JKT*) is a tree that satisfies the following conditions.

- Each node in the tree contains at least one keyword.

- The tree contains all the keywords (total), and there exist no subtrees that contain all the keywords (minimal).

- Each edge of the tree has a positive integer weight, and the total weight for all edges in the tree is smaller than Tmax.[1]

- For any two keywords w_i and w_j contained in nodes v_1 and v_2, respectively, suppose the distance (total weight of edges) between v_1 and v_2 in the tree is d, then there exists an edge (w_i, w_j) in G whose weight is d.

[1] Tmax is the maximum number of nodes allowed in a tree.

An example of a *KRG* is shown in Figure 5.1(a) where there are five keywords w_1, w_2, ..., w_5. For edge (w_1, w_5), the two keywords are contained in a certain tuple in database D, so we add weight 0. There also exists a joining sequence of length 2 that connects w_1 and w_5 at the two ends, so we add weight 2. A *JKT* of the *KRG* is shown in Figure 5.1(b). For two keywords w_1 and w_2 in the *JKT*, their distance in the tree is 0 because they are contained in the same node. The edge (w_2, w_3) of the *KRG* has weight 0. The distance of the two keywords w_2 and w_4 is 3, and we can also find that edge (w_2, w_4) has weight 3. Given a database D and its *KRG* G, we have the following theorem.

Theorem 5.1   *Given a keyword query Q, for a database $D \in \mathcal{D}$, if its KRG does not contain a JKT for Q, the results (MTJNTs) for the keyword query Q over the database D will be empty.*

Using Theorem 5.1, it can prune databases that do not contain a *JKT* for the keyword query Q. For other databases, a new scoring function is defined in order to rank them. The scoring function considers both the IR ranking score and the structural score (distances between keywords). The score consists of two parts, namely the node score and the edge score. For database D that is not pruned and for the keyword space $\mathcal{K}$, the node score and the edge score are as follows:

- **The node score:** The score of each keyword $w_i \in \mathcal{K}$ is

$$score(D, w_i) = \frac{\sum_{t \in D \text{ and } t \text{ contains } w_i} score(t, D, w_i)}{N(D, w_i)} \tag{5.6}$$

where $N(D, w_i)$ is the number of tuples in D that contain keyword w_i and the score for each tuple t with respect to w_i, $score(t, D, w_i)$ is defined as follows:

$$score(t, D, w_i) = \frac{tf(t, w_i)}{\sum_{w \in t} tf(t, w)} \cdot \ln \frac{N(D)}{N(D, w_i) + 1} \tag{5.7}$$

where $tf(t, w_i)$ is the term frequency of w_i in the tuple t, and $N(D)$ is the total number of tuples in D.

- **The edge score:** For any two keywords $w_i \in \mathcal{K}$ and $w_j \in \mathcal{K}$, the edge score $score(D, w_i, w_j)$ is defined as follows:

$$score(D, w_i, w_j) = \sum_{d=1}^{\delta} score_d(D, w_i, w_j) \tag{5.8}$$

Here δ is a parameter to control the maximum distance between two keywords, and

$$score_d(D, w_i, w_j) = \frac{\sum_{(t,t') \in P_d(w_i, w_j, D)} tf(t, w_i) \cdot tf(t', w_j) \cdot \ln \frac{N_d(D)}{N_d(w_i, w_j, D) + 1}}{N_d(w_i, w_j, D)} \tag{5.9}$$

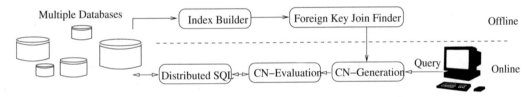

Figure 5.2: The architecture of *Kite*

where $P_d(w_i, w_j, D)$ is the set of tuple pairs defined as: $P_d(w_i, w_j, D) = \{(t, t')| t \in D, t' \in D, t$ contains w_i, t' contains w_j, t and t' can be joined in a sequence of length d in $D\}$. $N_d(D)$ is the total number of tuple pairs (t, t') in D such that t and t' can be joined in a sequence of length d. $N_d(w_i, w_j, D)$ is the total number of tuple pairs (t, t') in D such that t contains w_i, t' contains w_j, t and t' can be joined in a sequence of length d in D. $N_d(w_i, w_j, D) = |P_d(w_i, w_j, D)|$.

- **The final score:** Given the node and edge scores, for the keyword query $Q \subseteq \mathcal{K}$, the score of database $D \in \mathcal{D}$ is defined as:

$$score(D, Q) = \sum_{w_i \in Q, w_j \in Q, i < j} score(D, w_i) \cdot score(D, w_j) \cdot score(D, w_i, w_j) \quad (5.10)$$

The databases with the top-k scores computed this way are chosen to answer query Q.

5.1.2   ANSWERING KEYWORD QUERIES ACROSS DATABASES

Given the set of multiple databases to be evaluated, a distributed keyword query finds a set of *MTJNT*s such that the tuples in each *MTJNT* may come from a different database. In *Kite* [Sayyadian et al., 2007], a framework to answer such a distributed keyword query is developed (Figure 5.2). We discuss the main components below.

Foreign Key Join Finder: The foreign key join finder discovers the foreign key reference between tuples from different databases. For each pair of tables U and V in different databases, there are 4 steps to find the foreign key references from tuples in U to tuples in V.

1. Finding keys in table U. In this step, a set of key attributes are discovered to be joined in table V. The algorithms developed in *TANE* [Huhtala et al., 1999] are adopted.

2. Finding joinable attributes in table V. For the set of keys in U found in the first step, a set of attributes are found in table V that can be joined with these keys. The algorithm *Bellman* [Dasu et al., 2002] is used for this purpose.

3. Generating foreign key join candidates. In this step, all foreign key references are generated between tuples in U and V using the above found joinable attributes.

4. Removing semantically incorrect candidates. This can be done using the schema matching method introduced in *Simflood* [Melnik et al., 2002].

CN-Generation: After finding the foreign key joins among databases, the database schema of all databases can be considered as a large database schema including two parts of edges:(1) foreign key references for tables in the same database and (2) foreign key references for tables in different databases. In order to generate the set of *CN*s in the large integrated database schema, any *CN* generation algorithm introduced in Chapter 2 can be adopted. As the database schema can be very large, this method may generate an extremely large number of *CN*s, which is inefficient. In *Kite*, the authors proposed to generate only the "condensed" *CN*s as follows: (1) combine all parallel edges (edges connect the same two tables) in the integrated schema into one edge and generate a condensed schema, (2) generate *CN*s on the condensed schema. In this way, the number of *CN*s can be largely reduced.

CN-Evaluation: In *Kite*, the set of *CN*s are evaluated using an iterative refinement approach. Three refinement algorithms are proposed, namely, Full, Partial, and Deep. Full is an adaption of the iterative refinement algorithm Sparse as introduced in Chapter 2. Partial is an adaption of the iterative refinement algorithm Global-Pipelined as introduced in Chapter 2. Deep joins each new selected tuple to be evaluated with all tuples including the unseen tuples in the corresponding tables. This is in contrast to Partial, where for each new tuple to be evaluated it considers joins for the new tuple with all the seen tuples so far. This method may increase much cross-database-joining cost when posing distributed SQL queries. Deep, on the other hand, considerably reduces the number of distributed SQL queries.

5.2   KEYWORD SEARCH ON SPATIAL DATABASES

In the context of keyword search on spatial databases, a spatial database $D = \{o_1, o_2, ...\}$ is a collection of objects. Each object o, consists of two parts, $o.k$ and $o.p$, where $o.k$ is a string (a collection of keywords) denoting the text associated with o to be matched with keywords in the query, and $o.p = (o.p_1, o.p_2, ..., o.p_d)$ is a d-dimensional point, specifying the spatial information (location) of o. There are two types of queries for keyword search on spatial databases based on the nature of results, those who return individual points (objects) and those who return areas.

5.2.1   POINTS AS RESULT

In this case, the keyword query Q consists of two parts, a list of keywords $Q.k = (Q.k_1, Q.k_2, ..., Q.k_l)$, and a d-dimensional point $Q.p = (Q.p_1, Q.p_2, ..., Q.p_s)$ specifying the location of Q. Suppose that there is a ranking function $f(dis(Q.p, o.p), irscore(Q.k, o.k))$ for any object $o \in D$, where $dis(Q.p, o.p)$ is the high dimensional distance between $Q.p$ and $o.p$, $irscore(Q.k, o.k)$ is the IR relevance score of query $Q.k$ to text $o.k$, and f is a function decreasing with $dis(Q.p, o.p)$ and increasing with $irscore(Q.k, o.k)$. Given a spatial database D, keyword query Q, and the ranking function f, the top-k keyword query is to get the top-k objects from D such that the function f for each top-k object is no smaller than any other non-top-k objects.

There are two naive methods to solve such a problem. The first method is to use R-Tree to retrieve objects in increasing order of dis. Each time an object is retrieved, it can update the upper bound of the f function for all the unseen objects. Once the upper bound is no larger than the k-th largest score of all seen tuples, it can stop and output the top-k objects found so far. The second method is to use an inverted list to get objects in decreasing order of $irscore$ and use an approach similar to the first method to get the top-k objects.

In [Felipe et al., 2008], a new structure called IR^2-Tree is introduced. An IR^2-Tree is similar as an R tree to index objects in D. The only difference is that, in each entry M (including leaf nodes) of an IR^2-Tree, there is an additional signature $M.sig$, recording the set of keywords contained in all objects located in the block area of the entry. The signature can be any compressed data structure to save space (e.g., the bitmap or the multi-level superimposed codes). Using the signature information, when processing queries, it can retrieve entries in the IR^2-Tree in a depth first manner and each time an entry is retrieved. It adopts a branch and bound method as follows. It calculates the upper bound of dis and the lower bound of $irscore$ simultaneously for the visited entry, thereby calculating the upper bound of the f function for the entry. If the upper bound is no larger than the k-th largest f value found so far, the whole tree rooted at this entry can be eliminated.

5.2.2   AREA AS RESULT

In this case, a keyword query $Q = \{k_1, k_2, ..., k_l\}$ is a list of keywords, and an answer for the keyword query is the smallest d-dimensional circle c spanned by objects $o_1, o_2, ..., o_l$, denoted $c = [o_1, o_2, ..., o_l]$ ($o_i \in D$ for $1 \leq i \leq l$), such that o_i contains keyword k_i for all $1 \leq i \leq l$ (i.e., $k_i \in o_i.k$) and the diameter of c, $diam(c)$ is minimized. The diameter of $c = [o_1, o_2, ..., o_l]$ is defined as follows:

$$diam(c) = max_{o_i \in c, o_j \in c} dis(o_i.p, o_j.p) \tag{5.11}$$

where $dis(o_i.p, o_j.p)$ is the k-dimensional distance between points $o_i.p$ and $o_j.p$. An example of the keyword query results is shown in Figure 5.3, where each object has a two dimensional location and contains one of the keywords $\{k_1, k_2, k_3\}$. The result of query $Q = \{k_1, k_2, k_3\}$ is the circle shown in Figure 5.3.

In order to find the result, in [Zhang et al., 2009], a new structure called BR*-Tree is introduced. It is similar to an R-Tree that indexes all objects in D, the only difference is that, in each entry M (including leaf nodes) of the BR*-Tree, there are two additional structures, $M.bmp$ and $M.kwd\_mbr$. $M.bmp$ is a bitmap of keywords, each position i of $M.bmp$ is either 0 or 1, specifying whether the MBR (Minimum Bounding Rectangle) of the entry contains keyword w_i or not for all $w_i \in \mathcal{K}$ ($\mathcal{K}$ is the entire keyword space). $M.kwd\_mbr$ is the vector of keyword MBR for all the keywords contained in the entry. Each keyword MBR for keyword w_i is the minimum bounding rectangle that contains all w_i in the entry. An example of an entry of the BR*-Tree is shown in Figure 5.4.

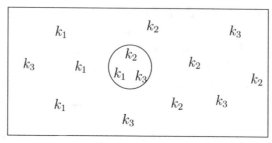

Figure 5.3: The result for the query $Q = \{k_1, k_2, k_3\}$

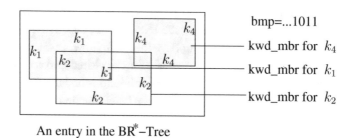

Figure 5.4: Illustration of an entry in the BR*-Tree

Given the BR*-Tree of a spatial database D and a keyword query $Q = \{k_1, k_2, ..., k_l\}$, the algorithm to search the minimal bounding circle $c = [c_1, c_2, ..., c_l]$ is as follows. It visits each entry in the BR*-Tree in a depth first fashion, and it keeps the minimal diameter among all circles found so far (d^*). For each new entry (or a set of new entries) visited, it enumerates all combinations of entries $C = (M_1, M_2, ..., M_s)$ such that each M_i is a sub-entry of a new entry and C contains all the keywords and $s \leq l$. If C has the potential to generate a better result, it decomposes C into a set of smaller combinations, and for each smaller combination, it recursively performs the previous steps until all entries in C are leaf nodes. In this situation, it uses the new result to update d^*. If C does not have the potential to generate a better result, it simply eliminates C and all combinations of the sub-entries generated from C. Here $C = (M_1, M_2, ..., M_s)$ has the potential to generate a better result iff it is distance mutex and keyword mutex which are defined below. Finally, it outputs the circle that has the diameter d^* as the final result of the query.

Definition 5.2 Distance Mutex. An entry combination $C = (M_1, M_2, ..., M_s)$ is distance mutex iff there are two entries $M_i \in C$ and $M_j \in C$ such that $dis(M_i, M_j) \leq d^*$. Here $dis(M_i, M_j)$ is the minimal distance between the MBR of M_i and the MBR of M_j.

Definition 5.3 Keyword Mutex.   An entry combination $C = (M_1, M_2, ..., M_s)$ is keyword mutex iff for any s different keywords in the query, $(k_{p_1}, k_{p_2}, ..., k_{p_s})$, where k_{p_i} is uniquely contributed by M_i, there always exist two different keywords k_{p_i} and k_{p_j} such that $dis(k_{p_i}, k_{p_j}) \leq d^*$. Here $dis(k_{p_i}, k_{p_j})$ is the minimal distance between the keyword MBR for k_{p_i} in M_i and the keyword MBR for k_{p_j} in M_j.

5.3   VARIATIONS OF KEYWORD SEARCH ON DATABASES

The approaches discussed in Chapter 2 and Chapter 3 aim at finding structures (trees or subgraphs) that connect all the user given keywords. There are also approaches that return various kinds of results according to different user requirements. In this section, we will introduce them one by one.

5.3.1   OBJECTS AS RESULTS

In *ObjectRank* [Balmin et al., 2004; Hristidis et al., 2008; Hwang et al., 2006], a relational database is modeled as a labeled weighted bi-directed graph $G_D(V, E)$, where each node $v \in V(G_D)$ is called an object and is associated with a list of attributes. Given a keyword query $Q = \{k_1, k_2, ..., k_l\}$, *ObjectRank* ranks objects according to their relevance to the query. The relevance of an object to a keyword query may come from two parts: (1) the object itself contains some keywords in some attributes, (2) the objects that are not far away in the sense of shortest distance on the graph contain the keywords and transfer their authorities to the object to be ranked. As an example, for the *DBLP* database graph shown in Figure 2.2, each paper tuple and author tuple can be considered as an object. For the keyword query $Q = \{XML\}$, the paper tuple p_3 can be considered as a good result, because (1) p_3 contains the keyword "XML" in its title, and (2) p_3 is cited by other papers (such as p_2) that contain the keyword "XML". The idea is borrowed from the PageRank in Google. First, for each edge $(v_i, v_j) \in E(G_D)$, a weight is assigned that we call the weight the authority transfer rate $\alpha(v_i, v_j)$, which is defined as follows:

$$\alpha(v_i, v_j) = \begin{cases} \frac{\alpha(\mathcal{L}(v_i), \mathcal{L}(v_j))}{outdeg(v_i, \mathcal{L}(v_j))}, & if\ outdeg(v_i, \mathcal{L}(v_j)) > 0 \\ 0, & if\ outdeg(v_i, \mathcal{L}(v_j)) = 0 \end{cases} \quad (5.12)$$

Here $\mathcal{L}(v)$ is the label of the node v. $\alpha(\mathcal{L}(v_i), \mathcal{L}(v_j))$ is the authority transfer rate of the schema edge $(\mathcal{L}(v_i), \mathcal{L}(v_j))$, which is predefined on the schema. $outdeg(v_i, \mathcal{L}(v_j))$ is the number of outgoing edges of v_i with the label $(\mathcal{L}(v_i), \mathcal{L}(v_j))$.

Suppose there are n nodes in G_D, i.e., $V(G_D) = \{v_1, v_2, ..., v_n\}$. A is an $n \times n$ transfer matrix, i.e., $A_{i,j} = \alpha(v_i, v_j)$ if there is an edge $(v_i, v_j) \in E(G_D)$, otherwise, $A_{i,j} = 0$. For each keyword w_k, let $s(k_i)$ be a base vector $s(k_i) = [s_0, s_1, ..., s_n]^T$, where $s_i = 1$ if v_i contains keyword k_i and otherwise $s_i = 0$. Let $e = [1, 1, ..., 1]^T$ be a vector of length n. The following are the ranking factors for each object $v_i \in V(G_D)$.

- **Global *ObjectRank*:** The global *ObjectRank* vector r^G is defined as follows:

$$r^G = dAr^G + (1-d)\frac{e}{n} \tag{5.13}$$

Here, d is a constant in $(0, 1)$ and r^G is a vector of size n, where $r^G[i]$ denotes the global *ObjectRank* for object v_i. The intuition behind the global *ObjectRank* is that an object is important if it is pointed by many other important objects. The global *ObjectRank* for each object is only related to the structure of the graph and is keyword independent.

- **Keyword-specific *ObjectRank*:** For each keyword k, the keyword-specific *ObjectRank* vector r^k is defined as follows:

$$r^k = dAr^k + (1-d)\frac{s(k)}{n(k)} \tag{5.14}$$

Here, $n(k)$ is the number of objects that contain the keyword k and r^k is a vector of size n, where $r^k[i]$ denotes the *ObjectRank* for object v_i with respect to keyword k. The intuition behind the keyword-specific *ObjectRank* is that an object is important if (1) it is pointed by many other important objects, or (2) it contains a specific keyword k or it is pointed by many other important objects that contain the specific keyword k.

- **Multiple-Keywords *ObjectRank*:** The keyword-specific *ObjectRank* only deals with one keyword. When there are multiple keywords, $Q = \{k_1, k_2, ..., k_l\}$, they are combined to get the Multiple-Keywords *ObjectRank* score r^Q depending on the different semantics. More specifically, under the AND semantics, it becomes

$$r^Q[i] = \prod_{k \in Q} r^k[i] \tag{5.15}$$

and under the OR semantics, it becomes

$$r^Q[i] = \sum_a r^{k_a}[i] - \sum_{a<b} r^{k_a}[i] \cdot r^{k_b}[i] + \sum_{a<b<c} r^{k_a}[i] \cdot r^{k_b}[i] \cdot r^{k_c}[i] - ... \tag{5.16}$$

- **The final *ObjectRank*:** The final object rank $r^{G,Q}[i]$ for each object v_i with respect to the keyword query Q is defined as follows:

$$r^{G,Q}[i] = r^Q[i] \cdot (r^G[i])^g \tag{5.17}$$

where g is a parameter denoting the importance of the global *ObjectRank* in the final rank.

Based on the final *ObjectRank*, the top-k objects are returned to answer a given keyword query.

Chakrabarti et al. [2006] studied the binary (many to many) relationships between objects and documents. For each object t, suppose there are n documents associated with it, denoted as $D_t = \{d_1, d_2, ..., d_n\}$. For a keyword query $Q = \{k_1, k_2, ..., k_l\}$, let the score for keyword k_i on document d_j be $DocScore(d_j, k_i)$. Two classes of score functions are given to rank objects:

- **Row-Marginal Class:** Such kinds of score functions first compute scores on the level of keywords, and then compute the final score on the level of documents.

$$score(t, Q) = agg_{d \in D_t}(comb_{k \in Q}(DocScore(d, k))) \tag{5.18}$$

- **Column-Marginal Class:** Such kinds of score functions first compute scores on the level of documents, and then compute the final score on the level of keywords.

$$score(t, Q) = comb_{k \in Q}(agg_{d \in D_t}(DocScore(d, k))) \tag{5.19}$$

Based on the two classes of score functions and the monotonicity of the aggregate/combine function, several early stop strategies can be designed to obtain the top-k objects.

5.3.2   SUBSPACES AS RESULTS

KDAP [Wu et al., 2007] focuses on keyword search in On-Line Analytical Processing (OLAP). Given a data warehouse where the schema contains both the fact table and the dimensions, a keyword query $Q = \{k_1, k_2, \cdots, k_l\}$ is processed in two steps.

1. In the first *differentiation* step, a set of subspaces is found, such that each subspace can be considered as a star that consists of a set of dimension-to-fact tuple paths. In each such path, at least one keyword is matched to the text attributes of the dimension tuple. Each star can be considered as an interpretation of the keyword query. A user needs to select one star which will be processed in the second step.

2. In the second *exploration* step, the star (subspace) selected in the first step will be further divided into smaller subspaces (dynamic facets) according to the set of dimensions that do not appear in the star. The criteria to choose dimensions to be divided is application dependent. For example, it can choose the dimension such that the aggregated measure of the subspace is largely different from other subspaces in the same level. After such partitioning, the roll-up and drill-down operations are also allowed to navigate the result space.

For example, consider a keyword query $Q = \{IBM, notebook\}$ against a fact table with three dimensions (brand, name, color) and a measure price. In the first step, the subspaces such as (brand=IBM,name=notebook X61), (brand=IBM,name=notebook T60), etc., are returned. Suppose the user chooses the subspace (brand=IBM,name=notebook X61); in the second step, the chosen subspace is further divided into smaller subspaces, such as (brand=IBM,name=notebook X61,color=white),   (brand=IBM,name=notebook X61,color=black), etc., and the price for each subspace is aggregated.

In *Agg-Keyword-Query* [Zhou and Pei, 2009], aggregate keyword search over a relation R is studied. Suppose a relation R consists of two parts of attributes: the dimension attributes $A_D = \{d_1, d_2, ..., d_n\}$ and the text attributes $A_T = \{a_1, a_2, ..., a_s\}$. The dimension attributes are used to represent subspaces and the text attributes include several texts where keyword search is allowed. For a tuple t, let $t.d_i (1 \leq i \leq n)$ denote the value in the dimension attribute d_i of t. A tuple t is said to contain keyword k if k is contained in one of the text attributes of t. For the subspace $c = (v_1, v_2, ..., v_n)$ that consists of n values, let $S(c)$ denote the set of tuples in R, where for each tuple t, $t.d_i = c.v_i$ for all $1 \leq i \leq n$, where $v_i = *(1 \leq i \leq n)$ serves as a wildcard to indicate that all values can be matched. Then,

$$S(c) = \{t | t \in R \text{ and } (t.d_i = c.v_i \text{ or } c.v_i = * \text{ for all } 1 \leq i \leq n)\} \tag{5.20}$$

Given a keyword query $Q = \{k_1, k_2, ..., k_l\}$, a subspace c is said to match Q if for all $k \in Q$, k is contained in at least one tuple in $S(c)$. For any two subspaces, c_1 and c_2, $c_1 \prec c_2$ if for all $1 \leq i \leq n$, if $c_2.v_i \neq *$ then $c_1.v_i = c_2.v_i$. The answer to keyword query Q consists of all subspaces, such that for each subspace c, (1) c matches Q (Total) and (2) there is no subspace c' such that c' also matches Q and $c' \prec c$ (Minimal). The answer can be generated by iteratively joining the list of tuples that contain keyword k_i with the list of tuples that contain keyword k_{i+1} for all $(1 \leq i \leq l-1)$ in order to get a set of candidate spaces and then prune non-minimal spaces after each join.

As an example, suppose the relation R contains three dimension attributes $A_D = \{\text{brand,name,color}\}$ and 1 text attribute $A_T = \{\text{complaint}\}$, and the keyword query $Q = \{\text{monitor,keyboard}\}$. The result will contain a list of subspaces such that all keywords in Q are contained in the at least one text attribute of a tuple in the returned subspace. Such as (brand=IBM,name=notebook X61,color=*),(brand=IBM,name=notebook T60,color=black), etc. Subspaces such as (brand=IBM,name=notebook T60,color=*) will not be returned if there exists (brand=IBM,name=notebook T61,color=black) already, because the latter is more specific.

5.3.3   TERMS AS RESULTS

Frequent co-occurring term (*FCT*) [Tao and Yu, 2009] focuses on finding frequent co-occurring terms for keyword search in a relational database. Given a keyword query $Q = \{k_1, k_2, ..., k_l\}$, the traditional methods return a set of *MTJNT*s $\mathcal{T}(Q) = \{T_1, T_2, ...\}$ as the result of the keyword query. In *FCT*, rather than returning a set of *MTJNT*s, a set of terms are returned. For example, in the *DBLP* database, for the keyword query $Q = \{\text{database,management,system}\}$, one of the co-occurring terms may be Jim Gray, which means that Jim Gray is an expert of database management system. For an *MTJNT* T, and a term w, let $count(T, w)$ denote the number of occurrences of w in the T. The frequency of a term w with respect to keyword query Q is defined as follows:

$$freq(Q, w) = \sum_{T \in \mathcal{T}(Q)} count(T, w) \tag{5.21}$$

Given a keyword query Q and an integer k, *FCT* search retrieves the top-k terms with highest frequency $freq(Q, w)$ such that $w \notin Q$.

A naive approach to answering a *FCT* query is to first evaluate the keyword query in order to generate all *MTJNT*s. Then for each term w, it counts the term frequency $count(T, w)$ for all $T \in \mathcal{T}(Q)$ before summarizing them together to get the top-k terms. An efficient approach is proposed in *FCT* to avoid enumerating all *MTJNT*s in order to get the results. Suppose the set of *CN*s for keyword query Q is $\mathcal{C}(Q) = \{C_1, C_2, ...\}$. For each $C \in \mathcal{C}(Q)$, let *MTJNT* (C) denote the set of *MTJNT*s generated by C, and $\mathcal{T}(Q) = \bigcup_{C \in \mathcal{C}(Q)} MTJNT\ (C)$. Then, it can evaluate all *CN*s $C \in \mathcal{C}(Q)$ individually, i.e.,

$$freq(Q, w) = \sum_{C \in \mathcal{C}(Q)} freq(C, w) \tag{5.22}$$

where $freq(C, w)$ is the frequency of w that occurs in any *MTJNT* of *MTJNT* (C), i.e.,

$$freq(C, w) = \sum_{T \in MTJNT\ (C)} count\ (T, w) \tag{5.23}$$

A *CN* C is said a star *CN* if it can find a root node R in C such that all the other nodes $\{R_1, R_2, ..., R_s\}$ in C are connected to R. A two step approach is used to evaluate a star *CN*. In the first step, it finds the tuple frequency for all tuples in the database that contain keywords in *MTJNT* (C). In the second step, it counts the term frequencies using the tuple frequencies calculated in the first step. A non-star *CN* C can be made as a star *CN* by joining some tuples and considering the joint relation as a single node in the *CN*. For example, *CN* $C = A\{Michelle\} \bowtie W \bowtie P \bowtie C \bowtie P\{XML\}$ is not a star *CN*, we can make it as a star *CN* by combining $R = C \bowtie P\{XML\}$ as a single node R to obtain $C = A\{Michelle\} \bowtie W \bowtie R$ which becomes a star *CN*.

DataClouds [Koutrika et al., 2009] finds a set of terms called data clouds, that are most relevant to the keyword query. The set of terms can be used to guide the users to refine their searches. In *DataClouds*, given a keyword query $Q = \{k_1, k_2, ..., k_l\}$ over a relational database D, where each result of Q is not an *MTJNT*, a subgraph of the database graph is centered at tuple t that includes nodes/edges within a certain distance of t. Each result can be uniquely identified by the center tuple t, denoted $G(t)$. The answer to keyword query Q is $ans(Q) = \{G(t)|t \in D$ and $G(t)$ contains all keywords in $Q\}$. Here, $G(t)$ contains a keyword k if k is contained in any tuple of $G(t)$. The score of each result $G(t)$ is defined as $score(G(t), Q) = \sum_{k \in Q} tf(G(t), k) \cdot idf(k)$. Here $tf(G(t), k)$ is the IR ranking score by considering each $G(t)$ as a virtual document. For each term $w \notin Q$, *DataClouds* considers three kinds of scoring functions to rank w with respect to Q.

- **Popularity-based:** This score is similar to the term frequency score as defined in *FCT* [Tao and Yu, 2009]:

$$score(Q, w) = \sum_{G(t) \in ans(Q)} freq(G(t), w) \tag{5.24}$$

Here $freq(G(t), w)$ is the number of occurrences of w in $G(t)$.

- **Relevance-based:** As different terms will have different importance, the IR scores are used as term weights:

$$score(Q, w) = \sum_{G(t) \in ans(Q)} tf(G(t), w) \cdot idf(w) \tag{5.25}$$

- **Query-dependence:** The relevance-based score only considers the importance of a term itself, without considering which result the term comes from. In some situations, the more important result $G(t)$ is, the higher would be the weights of the terms that come from $G(t)$. Thus,

$$score(Q, w) = \sum_{G(t) \in ans(Q)} (tf(G(t), w) \cdot idf(w)) \cdot score(G(t), Q) \qquad (5.26)$$

In *FCT*, each term is treated equally, and each result is also treated equally. Whenever a term appears in a result, it will contribute a unit score to the final score. The popularity-based score is similar to the score defined in *FCT*. In the relevance-based score, each result is treated equally, but each term is not treated equally. The contribution of a term in a result is proportional to its TF-IDF relevance score with respect to the result. In the query-dependence score, each result is not treated equally, and each term is not treated equally. The contribution of a term in a result is proportional to its TF-IDF relevance score as well as the query's TF-IDF relevance score with respect to the result.

5.3.4 SQL QUERIES AS RESULTS

SQL queries enable users to query relational databases, but also require users to have knowledge of RDBMS as well as the syntax of SQL. For a non-expert user, a practical and easy way to query a relational database is to use a keyword query that is less expressive than SQL. Therefore, there is a trade-off between the expressive power of the user given query and the ease to use the query.

Given a keyword query over a relational database, many systems (e.g., [Chu et al., 2009; Tata and Lohman, 2008; Tran et al., 2009]) attempt to interpret the keyword query into top-k SQL queries that can best explain the user given keyword query. Such SQL queries can be represented as forms [Chu et al., 2009] or conjunctive queries [Tran et al., 2009]. All such systems use similar ideas to generate SQL queries. We introduce *SQAK* [Tata and Lohman, 2008] as the representative approach.

In *SQAK*, the system enables users to post keyword queries that include aggregations. For example, for the *DBLP* database with schema graph shown in Figure 2.1, a user can post the keyword query $Q = \{author, number, paper, XML\}$ to get the number of papers about "XML" for each author. One of the possible results for such a query can be the following SQL:

select count(Paper.TID), Author.TID, Author.Name **from** Paper, Write, Author
where Paper.TID=Write.PID **and** Write.AID=Author.TID **and** contain(Paper.Title,XML)
group by Author.TID, Author.Name

Formally, a keyword query consists of two parts: the general keywords, such as "paper", "author", and "XML", and the aggregate keywords, such as "number". In the first step of the *SQAK* system, the keyword query is interpreted into a list of Candidate Interpretations (*CI*), where each $CI = (C, a, F, w)$ includes four parts: C includes a set of attributes and a prediction on some of the attributes (e.g., the keywords contained in each attribute), a is an attribute in C over which the aggregate function is performed, F is the aggregate function, and w is the group-by attribute in C

(maybe empty). For example, for the keyword query $Q = \{$author, number, paper, XML$\}$, one of the possible CIs is ($C = \{$author.TID, paper.TID, paper.title contains "XML" $\}$, $a =$ paper.TID, $F =$ count, $w =$ author.TID). A CI is trivial if one of the following is satisfied: (1) C contains a attribute $c \neq a$ such that c functionally determines a, or (2) C contains two attributes c_i and c_j that refer to the same attribute or c_i is a foreign key of c_j. The set of non-trivial CIs can be enumerated by using the full text index enabled in RDBMS.

After enumerating all non-trivial CIs, for each $CI = (C, a, F, w)$, it enumerates a set of Simple Query Networks (SQN) where each SQN is a connected subgraph of the schema graph that satisfies the following conditions:

- **Total** - All tables in C are contained in the SQN.

- **Minimal** - It is not total if any node is removed from SQN.

- **Node Clarity** - Each node in SQN has at most one incoming edge.

Suppose the cost of a SQN is the summation of all edge costs and node costs. For each CI, it needs to get the SQN with the smallest cost, which is a NP-Complete problem. A heuristic greedy algorithm is proposed in $SQAK$. For a CI, (C, a, F, w), it starts at the table o that contains the attribute a. For each of the other tables (nodes) $v \in C$, it finds the shortest path from v to o in a backtrack manner. If, after adding the path from v to o, the node clarity condition is violated, it backtracks to find the next shortest path from v to o until all nodes in C are successfully added. It then outputs the current result to be a good SQN for the CI. After finding the SQN for each CI, it can get the top-k SQNs with the smallest cost. And each of the top-k SQNs is translated into an SQL to be output.

5.3.5   SMALL DATABASE AS RESULT

Précis [Koutrika et al., 2006; Simitsis et al., 2008] returns a small database that contains only the tuples relevant to a given keyword query Q. The schema of a relational database D is modeled as a weighted graph $G_S(V, E)$, where each relation is modeled as a node in G_S, and each foreign key reference between relations is modeled as an edge in G_S. Each edge also has a weight, defining the tightness of the relationship between the two relations. Given a keyword query $Q = \{k_1, k_2, ..., k_l\}$, the result of applying Q on D is a small database D', satisfying the following conditions.

1. The set of relation names in D' is a subset of the set of relation names in D.

2. For each relation $R' \in D'$ that corresponds to relation $R \in D$, we have $att(R') \subseteq att(R)$ and $tup(R') \subseteq tup(R)$, where $att(R)$ denotes the attributes of R and $tup(R)$ denotes the tuples of R.

3. The tuples in D' can be generated by expanding from the tuples that contain keywords in the query, following the foreign key references. They must satisfy the degree constraints and

cardinality constraints. Degree constraints define the attributes and relations in D'. They include (1) the maximum number of attributes in D', and (2) the minimum weight of projection paths in the database schema graph G_S. Cardinality constraints define the set of tuples in D'. They include (1) the maximum number of tuples in D', and (2) the maximum number of tuples for each relation in D'.

For example, for the *DBLP* database shown in Figure 2.2, consider a keyword query $Q = \{algorithms\}$, with the constraint such that the distance from any tuple to the tuple that contains the keyword in Q must be no larger than 2. Then, the result contains the database having the same schema with the original database. Tuples such as p_2 and p_3 will be contained in the result because they all have distance 2 with the tuple p_4 that contains the keyword "algorithms". Tuples such as a_1, a_2 and p_4 will not be contained in the result because they all have distance larger than 2 with any tuple that contains the keyword "algorithms".

In *Précis*, a keyword query is processed in two steps. In the first step, the schema of the database D' is generated, such that all of the degree constraints are satisfied. This can be done easily by expanding from the relations, that may contain the user given keywords, to the adjacent relations following the foreign key references, until all degree constraints are satisfied. In the second step, it evaluates each join edge defined in the schema of D' in order to satisfy all the cardinality constraints.

5.3.6   OTHER RELATED ISSUES

Jagadish et al. [2007] assert that usability of a database is an important issue to address in database research. Enabling keyword query on database is one aspect to improve the usability.

Goldman et al. [1998] propose the notion, *proximity search*, which is to search objects in database that are "near" other relevant objects. Here the database is represented as a graph, where objects are represented by nodes and edges represent relationships between the corresponding objects.

Su and Widom [2005] propose to construct virtual documents offline, which is the answer unit for a keyword query. Virtual documents are interconnected tuples from multiple relations. Query answering is in an traditional IR style, where virtual documents satisfying the query are returned. Nandi and Jagadish [2009] propose to represent the database, conceptually, as a collection of independent "queried units", each of which represents the desired result of some query against the database. Jayapandian and Jagadish [2008] present an automated technique to generate a good set of forms that can express all possible queries, and each form is capable of expressing only a very limited range of queries. Talukdar et al. [2008] present a system with which a non-expert user can author new query templates and Web forms, to be used by anyone with related information needs. The query templates and Web forms are generated by a keyword query against interlinked source relations.

Ji et al. [2009] study interactive keyword search on *RDB*, where the interaction is provided by *autocompletion*, which predicts a word of phrase that a user may type based on the partial query the user has entered. An answer defined in [Ji et al., 2009] is a single record in *RDB*. Li et al. [2009a] extend the *autocompletion* framework to the steiner tree based semantics for a keyword query.

Chaudhuri and Kaushik [2009] study autocompletion with tolerated errors in a general framework, in which only autocompletions are computed without query evaluation. [Pu and Yu, 2008, 2009] study the problem of query cleaning for keyword queries in *RDB*, where query cleaning involves semantic linkage and spelling corrections followed by segmenting nearby query words into high quality data terms.

Guo et al. [2007] present efficient algorithm to conduct topology search over biological databases. Shao et al. [2009b] present an effective workflow search engine, WISE, to find informative and concise search results, defined as the minimal views of the most specific workflow hierarchies containing keywords for a keyword query.

Bibliography

Sanjay Agrawal, Surajit Chaudhuri, and Gautam Das. DBXplorer: A system for keyword-based search over relational databases. In *Proc. 18th Int. Conf. on Data Engineering*, pages 5–16, 2002. DOI: 10.1109/ICDE.2002.994693 **2.1, 2.3**

Shurug Al-Khalifa, Cong Yu, and H. V. Jagadish. Querying structured text in an xml database. In *Proc. 2003 ACM SIGMOD Int. Conf. On Management of Data*, pages 4–15, 2003. DOI: 10.1145/872757.872761 **4.5**

Sihem Amer-Yahia, Pat Case, Thomas Rölleke, Jayavel Shanmugasundaram, and Gerhard Weikum. Report on the db/ir panel at sigmod 2005. *SIGMOD Record*, 34(4):71–74, 2005. DOI: 10.1145/1107499.1107514 (document)

Sihem Amer-Yahia and Jayavel Shanmugasundaram. Xml full-text search: Challenges and opportunities. In *Proc. 31st Int. Conf. on Very Large Data Bases*, page 1368, 2005. (document)

Andrey Balmin, Vagelis Hristidis, Nick Koudas, Yannis Papakonstantinou, Divesh Srivastava, and Tianqiu Wang. A system for keyword proximity search on xml databases. In *Proc. 29th Int. Conf. on Very Large Data Bases*, pages 1069–1072, 2003. **4.5**

Andrey Balmin, Vagelis Hristidis, and Yannis Papakonstantinou. ObjectRank: Authority-based keyword search in databases. In *Proc. 30th Int. Conf. on Very Large Data Bases*, pages 564–575, 2004. **5.3.1**

Zhifeng Bao, Tok Wang Ling, Bo Chen, and Jiaheng Lu. Effective xml keyword search with relevance oriented ranking. In *Proc. 25th Int. Conf. on Data Engineering*, pages 517–528, 2009. DOI: 10.1109/ICDE.2009.16 **4.5**

Gaurav Bhalotia, Arvind Hulgeri, Charuta Nakhe, Soumen Chakrabarti, and S. Sudarshan. Keyword searching and browsing in databases using BANKS. In *Proc. 18th Int. Conf. on Data Engineering*, pages 431–440, 2002. DOI: 10.1109/ICDE.2002.994756 **3.1, 3.1, 3.3.1, 3.3.1**

Sergey Brin and Lawrence Page. The anatomy of a large-scale hypertextual web search engine. *Computer Networks*, 30(1-7):107–117, 1998. DOI: 10.1016/S0169-7552(98)00110-X **3.1, 4.4.1**

Kaushik Chakrabarti, Venkatesh Ganti, Jiawei Han, and Dong Xin. Ranking objects based on relationships. In *Proc. 2006 ACM SIGMOD Int. Conf. On Management of Data*, pages 371–382, 2006. DOI: 10.1145/1142473.1142516 **5.3.1**

Surajit Chaudhuri and Gautam Das. Keyword querying and ranking in databases. *Proc. of the VLDB Endowment*, 2(2):1658–1659, 2009. (document)

Surajit Chaudhuri and Raghav Kaushik. Extending autocompletion to tolerate errors. In *Proc. 2009 ACM SIGMOD Int. Conf. On Management of Data*, pages 707–718, 2009. DOI: 10.1145/1559845.1559919 5.3.6

Surajit Chaudhuri, Raghu Ramakrishnan, and Gerhard Weikum. Integrating db and ir technologies: What is the sound of one hand clapping? In *Proc. of CIDR'05*, 2005. (document)

Yi Chen, Wei Wang, Ziyang Liu, and Xuemin Lin. Keyword search on structured and semi-structured data. In *Proc. 2009 ACM SIGMOD Int. Conf. On Management of Data*, pages 1005–1010, 2009. DOI: 10.1145/1559845.1559966 (document)

Eric Chu, Akanksha Baid, Xiaoyong Chai, AnHai Doan, and Jeffrey F. Naughton. Combining keyword search and forms for ad hoc querying of databases. In *Proc. 2009 ACM SIGMOD Int. Conf. On Management of Data*, pages 349–360, 2009. DOI: 10.1145/1559845.1559883 5.3.4

Sara Cohen, Jonathan Mamou, Yaron Kanza, and Yehoshua Sagiv. XSEarch: A semantic search engine for XML. In *Proc. 29th Int. Conf. on Very Large Data Bases*, pages 45–56, 2003. 4.1.2, 4.5

Thomas H. Cormen, Clifford Stein, Ronald L. Rivest, and Charles E. Leiserson. *Introduction to Algorithms*. McGraw-Hill Higher Education, 2001. ISBN 0070131511. 3.2, 3.3.2

Bhavana Bharat Dalvi, Meghana Kshirsagar, and S. Sudarshan. Keyword search on external memory data graphs. *Proc. of the VLDB Endowment*, 1(1):1189–1204, 2008. DOI: 10.1145/1453856.1453982 2.4, 3.4.3, 3.4.3

Tamraparni Dasu, Theodore Johnson, S. Muthukrishnan, and Vladislav Shkapenyuk. Mining database structure; or, how to build a data quality browser. In *Proc. 2002 ACM SIGMOD Int. Conf. On Management of Data*, pages 240–251, 2002. DOI: 10.1145/564691.564719 2

Bolin Ding, Jeffrey Xu Yu, Shan Wang, Lu Qin, Xiao Zhang, and Xuemin Lin. Finding top-k min-cost connected trees in databases. In *Proc. 23rd Int. Conf. on Data Engineering*, pages 836–845, 2007. DOI: 10.1109/ICDE.2007.367929 3.1, 3.4, 3.3.2, 3.3.2

S. E. Dreyfus and R. A. Wagner. The steiner problem in graphs. In *Networks*, 1972. 3.1

Ronald Fagin. Fuzzy queries in multimedia database systems. In *Proc. 17th ACM SIGACT-SIGMOD-SIGART Symp. on Principles of Database Systems*, pages 1–10, 1998. DOI: 10.1145/275487.275488 3.5.1

Ronald Fagin, Amnon Lotem, and Moni Naor. Optimal aggregation algorithms for middleware. In *Proc. 20th ACM SIGACT-SIGMOD-SIGART Symp. on Principles of Database Systems*, 2001. DOI: 10.1145/375551.375567 3.2, 4.4.1

Ian De Felipe, Vagelis Hristidis, and Naphtali Rishe. Keyword search on spatial databases. In *Proc. 24th Int. Conf. on Data Engineering*, pages 656–665, 2008. DOI: 10.1109/ICDE.2008.4497474 5.2.1

Roy Goldman, Narayanan Shivakumar, Suresh Venkatasubramanian, and Hector Garcia-Molina. Proximity search in databases. In *Proc. 24th Int. Conf. on Very Large Data Bases*, pages 26–37, 1998. 5.3.6

Konstantin Golenberg, Benny Kimelfeld, and Yehoshua Sagiv. Keyword proximity search in complex data graphs. In *Proc. 2008 ACM SIGMOD Int. Conf. On Management of Data*, pages 927–940, 2008. DOI: 10.1145/1376616.1376708 3.1, 3.1, 3.2, 3.1, 3.3, 3.3.3, 3.3.3, 3.3.3, 3.3.3, 3.8, 3.10

Lin Guo, Jayavel Shanmugasundaram, and Golan Yona. Topology search over biological databases. In *Proc. 23rd Int. Conf. on Data Engineering*, pages 556–565, 2007. DOI: 10.1109/ICDE.2007.367901 5.3.6

Lin Guo, Feng Shao, Chavdar Botev, and Jayavel Shanmugasundaram. Xrank: Ranked keyword search over xml documents. In *Proc. 2003 ACM SIGMOD Int. Conf. On Management of Data*, pages 16–27, 2003. DOI: 10.1145/872757.872762 4.1.1, 4.2.2, 4.4.1, 4.4.1

Hao He, Haixun Wang, Jun Yang, and Philip S. Yu. BLINKS: ranked keyword searches on graphs. In *Proc. 2007 ACM SIGMOD Int. Conf. On Management of Data*, pages 305–316, 2007. DOI: 10.1145/1247480.1247516 2.4, 3.4.2, 3.4.2

Vagelis Hristidis, Luis Gravano, and Yannis Papakonstantinou. Efficient IR-Style keyword search over relational databases. In *Proc. 29th Int. Conf. on Very Large Data Bases*, pages 850–861, 2003a. 2.1, 2.3, 2.3.2

Vagelis Hristidis, Heasoo Hwang, and Yannis Papakonstantinou. Authority-based keyword search in databases. *ACM Trans. Database Syst.*, 33(1), 2008. DOI: 10.1145/1331904.1331905 2.4, 5.3.1

Vagelis Hristidis, Nick Koudas, Yannis Papakonstantinou, and Divesh Srivastava. Keyword proximity search in xml trees. *IEEE Trans. Knowl. and Data Eng.*, 18(4):525–539, 2006. DOI: 10.1109/TKDE.2006.1599390 4.5

Vagelis Hristidis and Yannis Papakonstantinou. DISCOVER: Keyword search in relational databases. In *Proc. 28th Int. Conf. on Very Large Data Bases*, pages 670–681, 2002. DOI: 10.1016/B978-155860869-6/50065-2 2.1, 2.2, 2.2, 2.3, 2.3.1

Vagelis Hristidis, Yannis Papakonstantinou, and Andrey Balmin. Keyword proximity search on xml graphs. In *Proc. 19th Int. Conf. on Data Engineering*, pages 367–378, 2003b. 4.5

Yu Huang, Ziyang Liu, and Yi Chen. extract: a snippet generation system for xml search. *Proc. of the VLDB Endowment*, 1(2):1392–1395, 2008a. 4.5

Yu Huang, Ziyang Liu, and Yi Chen. Query biased snippet generation in xml search. In *Proc. 2008 ACM SIGMOD Int. Conf. On Management of Data*, pages 315–326, 2008b. DOI: 10.1145/1376616.1376651 4.5

Ykä Huhtala, Juha Kärkkäinen, Pasi Porkka, and Hannu Toivonen. Tane: An efficient algorithm for discovering functional and approximate dependencies. *Comput. J.*, 42(2):100–111, 1999. DOI: 10.1093/comjnl/42.2.100 1

Heasoo Hwang, Vagelis Hristidis, and Yannis Papakonstantinou. Objectrank: a system for authority-based search on databases. In *Proc. 2006 ACM SIGMOD Int. Conf. On Management of Data*, pages 796–798, 2006. DOI: 10.1145/1142473.1142593 5.3.1

H. V. Jagadish, Adriane Chapman, Aaron Elkiss, Magesh Jayapandian, Yunyao Li, Arnab Nandi, and Cong Yu. Making database systems usable. In *Proc. 2007 ACM SIGMOD Int. Conf. On Management of Data*, pages 13–24, 2007. DOI: 10.1145/1247480.1247483 5.3.6

Magesh Jayapandian and H. V. Jagadish. Automated creation of a forms-based database query interface. *Proc. of the VLDB Endowment*, 1(1):695–709, 2008. DOI: 10.1145/1453856.1453932 5.3.6

Shengyue Ji, Guoliang Li, Chen Li, and Jianhua Feng. Efficient interactive fuzzy keyword search. In *Proc. 18th Int. World Wide Web Conf.*, pages 371–380, 2009. DOI: 10.1145/1526709.1526760 5.3.6

David S. Johnson, Christos H. Papadimitriou, and Mihalis Yannakakis. On generating all maximal independent sets. *Inf. Process. Lett.*, 27(3), 1988. DOI: 10.1016/0020-0190(88)90065-8 3.2

Varun Kacholia, Shashank Pandit, Soumen Chakrabarti, S. Sudarshan, Rushi Desai, and Hrishikesh Karambelkar. Bidirectional expansion for keyword search on graph databases. In *Proc. 31st Int. Conf. on Very Large Data Bases*, pages 505–516, 2005. 3.4.1

Beeny Kimelfeld and Yehoshua Sagiv. New algorithms for computing steiner trees for a fixed number of terminals. In *http://www.cs.huji.ac.il/ bennyk/papers/steiner06.pdf*, 2006a. 3.3.2

Benny Kimelfeld and Yehoshua Sagiv. Finding and approximating top-k answers in keyword proximity search. In *Proc. 25th ACM SIGACT-SIGMOD-SIGART Symp. on Principles of Database Systems*, pages 173–182, 2006b. DOI: 10.1145/1142351.1142377 3.1, 3.3, 3.3.3, 3.3.3, 3.5, 3.6, 3.3.3, 3.6, 3.7, 3.3.3, 3.8, 3.3.3, 3.9

Lingbo Kong, Rémi Gilleron, and Aurélien Lemay. Retrieving meaningful relaxed tightest fragments for xml keyword search. In *Advances in Database Technology, Proc. 12th Int. Conf. on Extending Database Technology*, pages 815–826, 2009. DOI: 10.1145/1516360.1516454 4.4.2, 4.4.2

Georgia Koutrika, Alkis Simitsis, and Yannis E. Ioannidis. Précis: The essence of a query answer. In *Proc. 22nd Int. Conf. on Data Engineering*, pages 69–78, 2006. DOI: 10.1109/ICDE.2006.114 5.3.5

Georgia Koutrika, Zahra Mohammadi Zadeh, and Hector Garcia-Molina. Data clouds: summarizing keyword search results over structured data. In *Advances in Database Technology, Proc. 12th Int. Conf. on Extending Database Technology*, pages 391–402, 2009. DOI: 10.1145/1516360.1516406 5.3.3

Eugene L. Lawler. A procedure for computing the k best solutions to discrete optimization problems and its application to the shortest path problem. *Management Science*, 18(7), 1972. 3.2, 3.5.2

Guoliang Li, Jianhua Feng, Jianyong Wang, and Lizhu Zhou. Effective keyword search for valuable lcas over xml documents. In *Proc. 16th ACM Conf. on Information and Knowledge Management*, pages 31–40, 2007a. DOI: 10.1145/1321440.1321447 4.1.1, 4.5

Guoliang Li, Shengyue Ji, Chen Li, and Jianhua Feng. Efficient type-ahead search on relational data: a tastier approach. In *Proc. 2009 ACM SIGMOD Int. Conf. On Management of Data*, pages 695–706, 2009a. DOI: 10.1145/1559845.1559918 5.3.6

Guoliang Li, Beng Chin Ooi, Jianhua Feng, Jianyong Wang, and Lizhu Zhou. EASE: an effective 3-in-1 keyword search method for unstructured, semi-structured and structured data. In *Proc. 2008 ACM SIGMOD Int. Conf. On Management of Data*, pages 903–914, 2008a. 2.4, 3.1, 3.5.1, 3.5.1, 3.15, 3.5.1

Guoliang Li, Xiaofang Zhou, Jianhua Feng, and Jianyong Wang. Progressive keyword search in relational databases. In *Proc. 25th Int. Conf. on Data Engineering*, pages 1183–1186, 2009b. DOI: 10.1109/ICDE.2009.196 3.4.2

Wen-Syan Li, K. Selçuk Candan, Quoc Vu, and Divyakant Agrawal. Retrieving and organizing web pages by "information unit". In *Proc. 10th Int. World Wide Web Conf.*, pages 230–244, 2001. DOI: 10.1145/371920.372057 3.1

Yunyao Li, Ishan Chaudhuri, Huahai Yang, Satinder Singh, and H. V. Jagadish. Danalix: a domain-adaptive natural language interface for querying xml. In *Proc. 2007 ACM SIGMOD Int. Conf. On Management of Data*, pages 1165–1168, 2007b. DOI: 10.1145/1247480.1247643 4.5

Yunyao Li, Cong Yu, and H. V. Jagadish. Schema-free xquery. In *Proc. 30th Int. Conf. on Very Large Data Bases*, pages 72–83, 2004. 4.1.2, 4.5

Yunyao Li, Cong Yu, and H. V. Jagadish. Enabling schema-free xquery with meaningful query focus. *VLDB J.*, 17(3):355–377, 2008b. DOI: 10.1007/s00778-006-0003-4 4.1.2, 4.5

Fang Liu, Clement T. Yu, Weiyi Meng, and Abdur Chowdhury. Effective keyword search in relational databases. In *Proc. 2006 ACM SIGMOD Int. Conf. On Management of Data*, pages 563–574, 2006. DOI: 10.1145/1142473.1142536 2.1, 2.1

Ziyang Liu and Yi Chen. Identifying meaningful return information for xml keyword search. In *Proc. 2007 ACM SIGMOD Int. Conf. On Management of Data*, pages 329–340, 2007. DOI: 10.1145/1247480.1247518 4.3, 4.3.1

Ziyang Liu and Yi Chen. Answering keyword queries on xml using materialized views. In *Proc. 24th Int. Conf. on Data Engineering*, pages 1501–1503, 2008a. DOI: 10.1109/ICDE.2008.4497603 4.5

Ziyang Liu and Yi Chen. Reasoning and identifying relevant matches for xml keyword search. *Proc. of the VLDB Endowment*, 1(1):921–932, 2008b. DOI: 10.1145/1453856.1453956 4.3, 4.5, 4.3.2, 4.8, 4.3.2, 4.3.2, 4.33, 4.4.2, 4.4.2

Ziyang Liu, Peng Sun, and Yi Chen. Structured search result differentiation. *Proc. of the VLDB Endowment*, 2(1):313–324, 2009a. 4.5

Ziyang Liu, Peng Sun, Yu Huang, Yichuan Cai, and Yi Chen. Challenges, techniques and directions in building xseek: an xml search engine. *IEEE Data Eng. Bull.*, 32(2):36–43, 2009b. 4.3.1

Ziyang Liu, Jeffrey Walker, and Yi Chen. Xseek: A semantic xml search engine using keywords. In *Proc. 33rd Int. Conf. on Very Large Data Bases*, pages 1330–1333, 2007. 4.3.1

Yi Luo, Xuemin Lin, Wei Wang, and Xiaofang Zhou. Spark: top-k keyword query in relational databases. In *Proc. 2007 ACM SIGMOD Int. Conf. On Management of Data*, pages 115–126, 2007. DOI: 10.1145/1247480.1247495 2.1, 2.1, 2.1, 2.1, 1, 2.3, 2.3.2, 2.3.2

Alexander Markowetz, Yin Yang, and Dimitris Papadias. Keyword search on relational data streams. In *Proc. 2007 ACM SIGMOD Int. Conf. On Management of Data*, pages 605–616, 2007. DOI: 10.1145/1247480.1247548 2.2, 2.2, 2.2, 2.2, 2.3

Sergey Melnik, Hector Garcia-Molina, and Erhard Rahm. Similarity flooding: A versatile graph matching algorithm and its application to schema matching. In *Proc. 18th Int. Conf. on Data Engineering*, pages 117–128, 2002. DOI: 10.1109/ICDE.2002.994702 4

Arnab Nandi and H. V. Jagadish. Qunits: queried units in database search. In *Proc. 4th Biennial Conf. on Innovative Data Systems Research*, 2009. 5.3.6

Ken Q. Pu and Xiaohui Yu. Keyword query cleaning. *Proc. of the VLDB Endowment*, 1(1):909–920, 2008. DOI: 10.1145/1453856.1453955 5.3.6

Ken Q. Pu and Xiaohui Yu. Frisk: Keyword query cleaning and processing in action. In *Proc. 25th Int. Conf. on Data Engineering*, pages 1531–1534, 2009. DOI: 10.1109/ICDE.2009.139 5.3.6

Lu Qin, Jeffrey Xu Yu, and Lijun Chang. Keyword search in databases: The power of rdbms. In *Proc. 2009 ACM SIGMOD Int. Conf. On Management of Data*, pages 681–694, 2009a. DOI: 10.1145/1559845.1559917 2.1, 2.2, 2.3, 2.3.1, 2.4

Lu Qin, Jeffrey Xu Yu, Lijun Chang, and Yufei Tao. Querying communities in relational databases. In *Proc. 25th Int. Conf. on Data Engineering*, pages 724–735, 2009b. DOI: 10.1109/ICDE.2009.67 2.4, 3.1, 3.5.2, 3.5.2, 3.5.2

Lu Qin, Jeffrey Xu Yu, Lijun Chang, and Yufei Tao. Scalable keyword search on large data streams. In *Proc. 25th Int. Conf. on Data Engineering*, pages 1199–1202, 2009c. DOI: 10.1109/ICDE.2009.200 2.3, 2.3.1

Mayssam Sayyadian, Hieu LeKhac, AnHai Doan, and Luis Gravano. Efficient keyword search across heterogeneous relational databases. In *Proc. 23rd Int. Conf. on Data Engineering*, pages 346–355, 2007. DOI: 10.1109/ICDE.2007.367880 2, 5.1.2

Feng Shao, Lin Guo, Chavdar Botev, Anand Bhaskar, Muthiah Chettiar, Fan Yang, and Jayavel Shanmugasundaram. Efficient keyword search over virtual xml views. In *Proc. 33rd Int. Conf. on Very Large Data Bases*, pages 1057–1068, 2007. 4.5

Feng Shao, Lin Guo, Chavdar Botev, Anand Bhaskar, Muthiah Chettiar, Fan Yang, and Jayavel Shanmugasundaram. Efficient keyword search over virtual xml views. *VLDB J.*, 18(2):543–570, 2009a. 4.5

Qihong Shao, Peng Sun, and Yi Chen. Wise: A workflow information search engine. In *Proc. 25th Int. Conf. on Data Engineering*, pages 1491–1494, 2009b. 5.3.6

Alkis Simitsis, Georgia Koutrika, and Yannis E. Ioannidis. Précis: from unstructured keywords as queries to structured databases as answers. *VLDB J.*, 17(1):117–149, 2008. DOI: 10.1007/s00778-007-0075-9 5.3.5

Qi Su and Jennifer Widom. Indexing relational database content offline for efficient keyword-based search. In *9th International Database Engineering and Applications Symposium*, pages 297–306, 2005. DOI: 10.1109/IDEAS.2005.36 5.3.6

Chong Sun, Chee Yong Chan, and Amit K. Goenka. Multiway slca-based keyword search in xml data. In *Proc. 16th Int. World Wide Web Conf.*, pages 1043–1052, 2007. DOI: 10.1145/1242572.1242713 4.2.2, 4.4, 4.2.2, 4.2.2

Partha Pratim Talukdar, Marie Jacob, Muhammad Salman Mehmood, Koby Crammer, Zachary G. Ives, Fernando Pereira, and Sudipto Guha. Learning to create data-integrating queries. *Proc. of the VLDB Endowment*, 1(1):785–796, 2008. DOI: 10.1145/1453856.1453941 5.3.6

Yufei Tao and Jeffrey Xu Yu. Finding frequent co-occurring terms in relational keyword search. In *Advances in Database Technology, Proc. 12th Int. Conf. on Extending Database Technology*, pages 839–850, 2009. DOI: 10.1145/1516360.1516456 5.3.3, 5.3.3

Sandeep Tata and Guy M. Lohman. SQAK: doing more with keywords. In *Proc. 2008 ACM SIGMOD Int. Conf. On Management of Data*, pages 889–902, 2008. DOI: 10.1145/1376616.1376705 5.3.4

Thanh Tran, Haofen Wang, Sebastian Rudolph, and Philipp Cimiano. Top-k exploration of query candidates for efficient keyword search on graph-shaped (rdf) data. In *Proc. 25th Int. Conf. on Data Engineering*, pages 405–416, 2009. DOI: 10.1109/ICDE.2009.119 5.3.4

Quang Hieu Vu, Beng Chin Ooi, Dimitris Papadias, and Anthony K. H. Tung. A graph method for keyword-based selection of the top-k databases. In *Proc. 2008 ACM SIGMOD Int. Conf. On Management of Data*, pages 915–926, 2008. DOI: 10.1145/1376616.1376707 1, 5.1.1

Ping Wu, Yannis Sismanis, and Berthold Reinwald. Towards keyword-driven analytical processing. In *Proc. 2007 ACM SIGMOD Int. Conf. On Management of Data*, pages 617–628, 2007. DOI: 10.1145/1247480.1247549 5.3.2

Yu Xu and Yannis Papakonstantinou. Efficient keyword search for smallest lcas in xml databases. In *Proc. 2005 ACM SIGMOD Int. Conf. On Management of Data*, pages 537–538, 2005. DOI: 10.1145/1066157.1066217 4.1, 4.1.1, 4.2.2, 4.2.2, 4.2.2, 4.2.2, 4.2.2

Yu Xu and Yannis Papakonstantinou. Efficient lca based keyword search in xml data. In *Proc. 16th ACM Conf. on Information and Knowledge Management*, pages 1007–1010, 2007. DOI: 10.1145/1321440.1321597 4.4.1

Yu Xu and Yannis Papakonstantinou. Efficient lca based keyword search in xml data. In *Advances in Database Technology, Proc. 11th Int. Conf. on Extending Database Technology*, pages 535–546, 2008. DOI: 10.1145/1353343.1353408 4.8, 4.4.1, 4.9, 4.4.1

Jin Y. Yen. Finding the k shortest loopless paths in a network. *Management Science*, 17(11):712–716, 1971. DOI: 10.1287/mnsc.17.11.712 3.2

Bei Yu, Guoliang Li, Karen R. Sollins, and Anthony K. H. Tung. Effective keyword-based selection of relational databases. In *Proc. 2007 ACM SIGMOD Int. Conf. On Management of Data*, pages 139–150, 2007. DOI: 10.1145/1247480.1247498 1, 5.1.1

Cong Yu and H. V. Jagadish. Schema summarization. In *Proc. 32nd Int. Conf. on Very Large Data Bases*, pages 319–330, 2006. 4.3.1

Dongxiang Zhang, Yeow Meng Chee, Anirban Mondal, Anthony K. H. Tung, and Masaru Kitsuregawa. Keyword search in spatial databases: Towards searching by document. In *Proc. 25th Int. Conf. on Data Engineering*, pages 688–699, 2009. DOI: 10.1109/ICDE.2009.77 5.2.2

Bin Zhou and Jian Pei. Answering aggregate keyword queries on relational databases using minimal group-bys. In *Advances in Database Technology, Proc. 12th Int. Conf. on Extending Database Technology*, pages 108–119, 2009. DOI: 10.1145/1516360.1516374 5.3.2

Authors' Biographies

JEFFREY XU YU

Jeffrey Xu Yu received his B.E., M.E., and Ph.D. in computer science, from the University of Tsukuba, Japan, in 1985, 1987 and 1990, respectively. Dr. Yu held teaching positions in the Institute of Information Sciences and Electronics, University of Tsukuba, Japan, and the Department of Computer Science, The Australian National University. Currently, he is a Professor in the Department of Systems Engineering and Engineering Management, the Chinese University of Hong Kong. Dr. Yu served as an associate editor of IEEE Transactions on Knowledge and Data Engineering, and is serving as a VLDB Journal editorial board member, and an Information Director of ACM Special Interest Group on Management of Data. His current main research interest includes keyword search in relational databases, graph database, graph mining, XML database, and Web-technology. He has published over 190 papers including papers published in reputed journals and major international conferences.

LU QIN

Lu Qin received the B.S. degree from Renmin University of China in 2006. He is currently a PhD candidate in the Department of Systems Engineering and Engineering Management, The Chinese University of Hong Kong, Hong Kong. His research interests include keyword search in relational databases, keyword search in large graphs and multi-query optimization in relational database management systems.

LIJUN CHANG

Lijun Chang received the B.S. degree from Renmin University of China in 2007. He is currently a PhD candidate in the Department of Systems Engineering and Engineering Management, The Chinese University of Hong Kong, Hong Kong. His research interests include uncertain data management and keyword search in databases.

Printed in the United States
by Baker & Taylor Publisher Services